Thiago Nascimento

The evolution of taxonomy

Thiago Nascimento

The evolution of taxonomy

Using genomes to identify and classify living beings

ScienciaScripts

Cover image: www.ingimage.com

This book is a translation from the original published under ISBN 978-613-9-63437-8.

Publisher:
Sciencia Scripts
is a trademark of
Dodo Books Indian Ocean Ltd. and OmniScriptum S.R.L publishing group

120 High Road, East Finchley, London, N2 9ED, United Kingdom
Str. Armeneasca 28/1, office 1, Chisinau MD-2012, Republic of Moldova, Europe
Printed at: see last page
ISBN: 978-620-7-98460-2

ACKNOWLEDGEMENTS

I would like to thank God first and foremost, because he has sustained me throughout my life.

To my beloved wife Ana Paula, who has always been by my side in happy times and difficult times, I thank you for your patience and support.

To my family who have always supported me and believed in me and my projects. To my mum Helena, who always worried about my future and supported me in everything I did.

To my father José Cícero, who has always been by my side at every important moment in my life, and to my siblings Thálisson, Thalita and Thaís, who always support and encourage me.

To all the teachers who were part of my education at UNEAL - Campus III, especially Prof Dr Deyvson Rodrigues Cavalcanti who always encouraged me and guided me with great dedication.

SUMMARY

The first attempts to organise living beings can be found in the works of Homer, Hippocrates and Herodotus, but Aristotle, in his day, was the one who produced the most extensive material on the subject. Decade after decade, knowledge about living beings increased, and the need for a system to identify and classify living beings became a crucial point for the advancement of science. Various authors contributed to Lineus finally being able to produce a classification system whose structure is still used today. With advances in research and the use of products derived from living organisms in areas such as pharmacology, ethnobotany, industry, among others, precise identification has become a necessity. The dichotomous keys commonly used to identify species are not always useful, because depending on the character analysed or the state of the sample, recognition becomes difficult or even impossible. With the evolution of molecular techniques and computer advances, molecular diagnosis has become part of research involving the classification of living beings. Various studies have been carried out using *software* capable of analysing sequences that make up the genome of living beings, making it possible to determine genetic similarities and establish kinship relationships. Areas such as phylogenetics have been the scene of much discussion and controversy, since analyses of kinship between species based on genes did not always corroborate with the phylogenies created from anatomical studies. So, with the intention of making the process of identifying species more practical, *DNA* barcoding emerged, similar to the barcodes used on products on market shelves. The *DNA barcode* is based on a short DNA sequence from the species' genome, which is enough to discriminate and classify it. To determine the target genes for DNA barcoding, various studies have been carried out to determine the most favourable genes, *COI* for animals, *matK* and *rbcL* in the *Plantae* kingdom, *ITS* for fungi and some intergenic stretches to determine species in the Monera kingdom. Once the best genes had been defined, the task became to sequence and store genes that could act as a *DNA barcode* for living beings. To this end, *BOLD Systems* was created, a system capable of supporting all phases of the analytical process of collecting *DNA barcode* samples. *BOLD* has more than five million specimen records and has provided support to research groups all over the world. The aim of this paper is to provide a bibliographical review of the basic concepts and current state of research and methods for classifying and identifying living beings based on their genomes.

Keywords: Systematics. Phylogeny. Bioinformatics. *DNA barcode.*

CONTENTS

INTRODUCTION

Since the dawn of civilisation, people have felt the need to get to know animals and plants. For many generations, products derived mainly from plants and animals have been used for the most diverse purposes: therapeutics (HAMILTON, 2004; LORENZI & MATOS, 2008), paint production (RODRIGUES, 2013), hallucinogens (MARTINEZ, 2009), poisons (SCHVARTSMAN, 1992), insecticides (MEDEIROS, 2005), cookery (LORENZI, 1994), among other applications. In order to guarantee the safety and efficacy of these natural products, it was necessary to identify and organise the raw material, which is living beings.

Organising beings was a task considered by many to be worthy of a monumental effort. There were hardly any instruments or techniques for this purpose at the time when Aristotle began producing the first materials on the classification of living beings (RIDLEY, 2009).

Decade after decade of accumulated knowledge generated various proposals on how to organise beings. Simple classifications, such as dividing them according to where they lived, were insufficient and inadequate because they grouped very different beings together. Knowledge up until the time of Aristotle considered mythology for the development of science (NASCIMENTO, 2011).

Even in publications well after Aristotle's time, it was common to see classifications with groups that included celestial beings, such as the diagram created by Ramom Llull (1232-1316) in 1305, entitled *"The Ladder of Ascent and Descent of the Intellect"* (Appendix A). In the diagram, Ramom Llull represents an evolutionary scale grouping, respectively: stones, fire, plants, animals, humans, heaven, angels and God.

Beliefs based on mythology met the need for answers to natural phenomena; however, these beliefs became dissociated from science as new knowledge about living beings was elucidated (RODRIGUES, 2016).

In the 15th and 16th centuries, the great navigations made it possible to explore previously unknown territories and, as a result, there was a considerable increase in the number of known species, making it clear that there was a need for increasingly effective mechanisms to name and organise living beings (CALDEIRA, 2009). Natural history museums were filled with specimens and catalogues and manuals were created to explain all the knowledge that was being acquired.

For many years, the organisation of living beings was based solely on the observation of morphological characters, groupings of living beings were made according to similarities between species, since the idea that beings were immutable persisted for a long time (RIDLEY, 2009). This idea became inconsistent with the consolidation of the results of Charles Darwin's (1809-1882) work on the evolution of species (RIDLEY, 2009).

New discoveries about evolution have led classifications to consider kinship relationships when grouping living beings, which could be done through morphological and chemical analyses and, more recently, gene analysis (RIDLEY, 2009).

In addition to the possibility of more reliable groupings of living beings (RUSSO, 2001), gene analysis points to new ways of identifying species and distinguishing between those that are apparently similar (HAJIBABAEI, 2006a; DE OLIVEIRA, 2013). Plant species, for example, can be difficult to identify if only morphological characteristics are considered, mainly from fragments of bark or leaves (CAVALCANTI & ALBUQUERQUE, 2013).

In this context, various software programmes associated with knowledge of the genome of living beings have made it possible to create new ways of classifying and identifying species. Among the forms of identification and classification is the emergence of a *DNA barcode* identification platform (HERBET, 2003), which promises to speed up the process of identifying living beings.

The aim of this paper is to provide a bibliographical review of the basic concepts and state of the art regarding methodologies for classifying and identifying living beings based on their genomes.

Chapter 2 of this work covers a historical review of classification systems, outlining the works and contributions of important researchers in the study of living beings. Later, in chapter 3, we'll look at the forms of classification based on the genome of living beings, how the main *software* for analysing DNA sequences works and discussions about the best genes for *DNA barcoding.*

Chapter 4 presents the applications and importance of *DNA barcoding* in some areas of science.

CHAPTER 1

ORIGIN OF CLASSIFICATION SYSTEMS

The first classifications of living beings appear in the works of Homer, Hippocrates and Herodotus. However, it was Aristotle who produced the most extensive material on the subject. The classification proposed by Aristotle was quite simple and took into account superficial characteristics of living beings, even if at some stage of the being's life such a characteristic was no longer visible.

The organisation that Aristotle proposed separated beings according to where they lived (terrestrial, aquatic, aerial, etc.) or how they reproduced. He used the principle of dichotomous classification, the development of which is attributed to Plato (427-347 BC). His studies left aside the mention of living beings as symbols of mythology and the cult of the gods, emphasising the empirical nature of the classification system.

However, the main difficulties pointed out by Aristotle were associated with plants, because plants vary much more than animals and often have periodic parts that can disappear at certain times, making identification more difficult, and in some cases impossible (ABREU, 1994). Aristotle believed that the world was immutable and, for this reason, grouping species by simple characteristics in common would be enough (PRESTES, 2009).

Many other philosophers and naturalists were concerned with the method developed by Aristotle. Among them, Theophrastus (362-287 BC) is a name that deserves to be highlighted when it comes to classification systems. Theophrastus was a disciple of Aristotle and developed some of the philosopher's ideas in the field of botany. Some terminology related to botany is attributed to Theophrastus, such as the terms root, stem, peduncle and fruit. Theophrastus also categorised plants according to stature: trees, shrubs, subshrubs and herbs. In addition, he established the difference between what we now call monocots and dicots and described floral parts in detail, even suggesting a division between flowering and non-flowering plants. As there were around 500 known plant species at the time, Theophrastus categorised them in their entirety (PRESTES, 2009).

Still in antiquity, another outstanding work was *Matéria Medica* (Figure 1) written by the Greek physician Dioscorides (60 AD). Dioscorides' main interest was to organise plants for medical purposes. *Matéria Medica* organised around 500 to 600 species of plants according to the products they produced, such as oils, resins, spices and fruits. Dioscorides criticised the method of listing plants in alphabetical order, saying that plants should be grouped according to common properties.

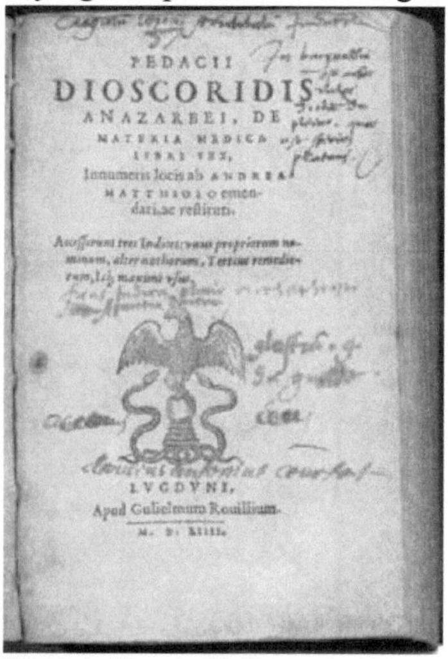

Figure 1: *Materia Medica*
Source: Historical Collections at the Claude Moore Health Sciences Library[1]

The works of Aristotle, Theophrastus and Dioscorides served as the basis for the study of the classification of living beings for more than fifteen hundred years, but were only really taken up again in the Renaissance period due to the need to organise an ever-increasing number of species. At this time, the number of known species increased considerably due to the frequency of great navigations, which explored previously unknown territories (CALDEIRA, 2009). The discoveries, especially those related to tropical regions, where the fauna and flora were more diverse, fuelled the

[1] Available at: <http://www.hsl.virginia.edu/historical/rare_books/herbalism/vienna.cfm> Accessed on 19 Aug. 2015

emergence of what would later become natural history museums dating back to the 16th century.

During the 16th century and the first half of the 17th century, the collections in museums began to have a more scientific slant, serving as a support for demonstrating, studying and disseminating knowledge about species (MARANDINO, 2009). The challenge was to record the species and organise them efficiently.

The search for a suitable method to classify species had been fuelled by the works of the Italian naturalist, botanist, doctor and philosopher Andrea Cesalpino (1519 - 1603). Cesalpino was the forerunner in the study of plant habits, types of fruit and seeds. He published *De Plantis* in 1583, in which he described around 1,500 plant species. In his work, Cesalpino denied the existence of sex in flowers (PANTOJA, 2007). To carry out the classification, Cesalpino had observed different groups of plants numerous times and concluded that plants could be grouped according to the characteristics of the leaves, stem and root, but only in the lower groups (herbaceous plants).

The upper groups (made up of woody plants), on the other hand, would be classified according to the fruit and parts involved in fruiting. Cesalpino also described the characteristics of the fruit, seeds and embryo in different groups of plants.

After Cesalpino, it was the physician and naturalist Jean Bauhin (1541-1613) who stood out in the history of the classification of living things, with a posthumous publication: *Historia Plantaram Universalis.* The focus of his work was the study of the texture and shape of leaves. *Historia Plantaram Universalis* was published in 1650 in Switzerland in two volumes, containing a record of 5,000 plants and 3,600 figures (VON MURALT, 2006).

In the 17th century, the Englishman John Ray (1627-1705) made important contributions to the study of plants. In 1660, John Ray published a book on the flora of Cambridge, which led to the publication of his *Catalogue of the Plants of England and the Adjacent Islands* ten years later. Ray travelled around studying the plants of various regions of Europe, as well as receiving specimens brought back by naturalists from America, Africa, India, China and Japan. Ray's most important work was *Historia plantaram generallis,* published in three volumes between 1686 and 1704, which catalogued around 18,000 plant species (PRESTES, 2009).

As well as inventorying various plants, John Ray also proposed a classification system based on a large number of characters. He was concerned with describing plants in a broad way, taking into account morphological characteristics, habitat, flowering time, life cycle (annual or perennial) and medicinal properties known at the time (PRESTES, 2009). Ray maintained the division between trees, shrubs and herbs previously proposed by Theophrastus. John Ray's greatest contributions were recognising the embryo and cotyledons, based on their external shape.

Figure 2: *Éléments de Botaniques,* written by Joseph Pitton de Tournefort. Source: Rare and Old Books[2]

In 1694, the Frenchman Joseph Pitton de Toumefort (1656-1708) appeared on the botanical scene with his work entitled *Éléments de botaniques* (Figure 2). In this publication, Toumefort classified plants based on the characters of the flower, more specifically the corolla, organising plants into groups and subgroups (PRESTES, 2009).

Toumefort was the first to use the term genus, which he considered to be the central concept for biological taxonomy, establishing 698 genera (RAVEN, 1971). Toumefort's method was very successful and was adopted by most naturalists at the time. Although the grouping was based on a small number of characters, this mechanism was used until the middle of the 16th centuryIL

[2] Available at: <http://cdn.livro-antigo.com/book/0000000010207/10207_1.jpg> Accessed on: 19 Aug. 2015

Since the 17th century, plant studies have involved a lot of observation, allowing scientists at the time to describe the behaviour of plants at different stages of development, which included reproduction. The idea that plants had a sex dates back to Greek antiquity, with Aristotle, who admitted the existence of two distinct sexes, but rejected the fact that they could be combined in the same individual. The English naturalist Nehemiah Grew (1628-1711) and the Italian Marcello Malpighi (1628-94), pioneers in the study of the anatomy of wood, recorded in their works in 1671 that most flowers had stamens and that this organ seemed to correspond to those that produced sperm fluids in animals (PRESTES, 2009).

Another researcher of the time who contributed to the discovery of sex in plants was John Ray, who realised that certain trees produce a greater quantity of fruit when female plants are sprayed with pollen from male plants. Ray himself called the globules released by the anthers of the stamens pollen.

However, ideas about the sexual characters present in plants were not very significant until the studies of Sébastien Vaillant (1669-1722). Vaillant analogised plant parts to human anatomy, thus proposing the presence of sexual organs in plants (WILLIAMS, 2001).

Experimental evidence about plant reproduction was subsequently described by the German botanist Rudolf Jacob Camerer (1675-1721). Camerer based his conclusions and experiments on careful observations and described the stamen as the male organ and the ovary as the female organ, emphasising the relationship with seed formation. Following in the same vein as John Ray's work, Camerer also described that most flowers had stamens and pistils, so that they self-fertilised and were therefore monoecious.

A large part of the theories and studies from this period served as the basis for the Swedish Karl von Linné (1707-1778), known as Linnaeus (the Portuguese version of his name) to develop his studies on the classification systems of living beings.

1.1 LINEUS' CONTRIBUTIONS TO BOTANY

From an early age, Linnaeus showed a certain tendency to study plants. Lineu began his studies around 1727, at the age of 20, at the University of Lund, where he was studying medicine in southern Sweden. The following year he was transferred to Uppsala University. Both universities were in quiet towns and had professors to teach rudimentary Lutheran orthodoxy. At the time, Linnaeus was aware of the classification system proposed by Toumefort and Sébastien Vaillant's sexual theory of plants, which stated that the stamens were comparable to the male sexual organs of animals and proposed that the terminology used for the floral parts should reflect their functions (PRESTES, 2009).

Most of the time Lineu was self-taught and, between 1728 and 1731, he worked as a guide in the botanical garden in Uppsala, as a tutor in the home of a professor of medicine and as a plant collector for a professor of theology, preparing a "Botanical Bible", as well as researching conifers and shrubs (JARDINE, 1996). Lineu wrote three plant catalogues at the botanical garden in Uppsala, using the systematic classification adopted at the time, proposed by Tournefort. In 1731, he wrote a new catalogue for the botanical garden in Uppsala. It was in this work that he changed Tournefort's systematics and adopted his own *"methodus própria"* or sexual classification system. Although he drew on Vaillant's work to propose his own classification system, he also incorporated contributions from other scholars, such as Tournefort, Cesalpino and Aristotle.

Linnaeus created a global classification system that was meant to cover all existing life forms. The classification system divided living beings into five levels of generality: class, order, genus, species and variety. He favoured the genus, a concept previously established by Tournefort, and subdivided the genus of plants according to the number, size, location and shape of the stamens and pistils. Linnaeus' artificial system divided the plant kingdom into 24 classes, with the first division being phanerogams (plants with visible sexual organs) and cryptogams (whose sexual organs are not visible). He also did this for animals, but they were grouped by a wide variety of characteristics, such as teeth, locomotion, blood type and habitat.

Figure 3: *Flora Lapponica,* written by Linnaeus in 1737.
Source: ADA online[3]

Lineu's central innovation was the classification based on the sexual parts of plants (JARDINE, 1996). By 1735, the essence of his systematic theory was complete. The system proposed by Linnaeus was applied for the first time in *Flora Lapponica* (Figure 3), from 1737, and *Hortas Cliffortianus,* written in the same year.

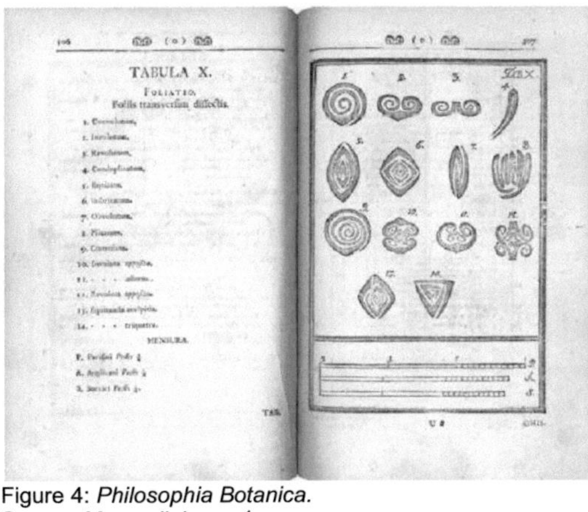

Figure 4: *Philosophia Botanica.*
Source: Meemelink.com[4]

Linnaeus also developed practical manuals to teach botany to novices.
In *Philosophia botanica* (1751), his most important guide, Linnaeus writes rules for botanical science. The book has entire pages with drawings of plant parts (Figure 4), as well as several indexes, and ends with a series of short texts in which he teaches how to create and manage a herbarium, as well as how to organise excursions for new

[3] Available at: <http://www.ada.auckland.ac.nz/4304floralapponica.jpg> Accessed on: 19 Aug. 2015.
[4] Available at: <http://www.meemelink.com/books_pages/27273.Linnaeus.htm> Accessed on: 19 Aug. 2015

discoveries.

In the same year that *Philosophia Botanica* was published, Linnaeus was summoned to Drottningholm Palace by Queen Lovisa Ulrika to describe her zoological collections. He also catalogued the zoological collections of King Adolf Fredrik, which was the first form of publication in which Linnaeus used binomial nomenclature (BLUNT, 2004).

The first major work in which Linnaeus used his system of naming living beings, called the binomial system, was *Species Plantarum,* in 1753 (SINGH, 2008). In this publication, Linnaeus defined that, in the binomial system, beings would be named by two Latin or Latinised words, the first word representing the genus of the living being and the second corresponding to the epithet, which together, genus and epithet, came to make up the scientific denomination of the species. Linnaeus also defined that scientific names should be printed in italics and when written by hand, they should be underlined. In the binomial system, only the first name (genus) should be written with the first letter capitalised, the rest in lower case.

Even after having published numerous important works *(Hortus Uplandicus, Systema Naturae, Genera Plantarum and Flora Laponica, Species Plantarum,* among others) and having created the most relevant artificial system, Linnaeus pointed out the need to create a natural system, one that was based on a greater number of characters present in living beings (PANTOJA, 2007).

For Linnaeus, an artificial classification system was one based on a single character, chosen by the author of the system, while the natural classification system was characterised by grouping beings based on various apparent characters (PRESTES, 2009).

Today, however, the concepts of artificial and natural classification systems are more complex than the one set out by Linnaeus in 1738. The artificial classification system is one in which classification takes place on the basis of arbitrary characters of the species, while the natural classification system groups living beings according to a set of relevant characters that allow them to be related evolutionarily.

As well as formalising binomial nomenclature, Linnaeus also established the classification of living beings in a hierarchy, using taxa, which are units of classification.

1.2 THE POST-LINEAR CLASSIFICATION

In the middle of the 18th century, a new phase in the study of plants was consolidated. With the increase in the number of plants collected and in need of identification, it was realised that there was a much greater natural affinity between plants than the Sexual System proposed by Linnaeus, in which only the flower was the organ needed to provide classification criteria (PRESTES, 2009). From this moment on, the scholars of the time began to propose systems based on various plant characters.

The Frenchman Michel Adanson (1727-1806) was one of the biggest critics of the system proposed by Lineu. After spending six years studying tropical flora in Senegal, Adanson was convinced that the systems proposed by Lineu were inadequate (FRANGSMYR & HEILBRON, 1990). Adanson was the first botanist to consider all plant parts together, proposing the "natural method of classification", in which he stated that each character could be studied and characterised according to quantity, image, relative proportion and symmetry.

FAMILLES
DES
PLANTES.

Par M. ADANSON, de l'Académie des Sciences, de la Société Roiale de Londres, Censeur Roial.

I. PARTIE.
Contenant une Préface Istorike sur l'état ancien & actuel de la Botanike, & une Téorie de cette Science.

Tot generibus Erbarum, utilitatibus hominum aut voluptatibus genitis recensitis, quantò plura restant, quantò que mirabiliora inventu! PLIN, Hist. nat. Lib. 22 Proœm.

A PARIS,
Chez VINCENT, Imprimeur-Libraire de Mgr le Comte de PROVENCE, rue S. Severin.

M DCC LXIII.
AVEC APPROBATION, ET PRIVILEGE DU ROI.

Figure 5: *Familles des Plantes* Source: Gallica Bibliothèque Numérique[5]

This first natural system was published in *Familles des plantes* (Figure 5) in 1763. The book set out the principles of the natural method applied to fifty-eight plant families. Among these principles were: the use of all possible plant characters, the determination of the singularities of each plant and the complete removal of similar characters between different species.

Adanson's project was considered by him to be comprehensive, or as he put it: "universal" - because he was concerned with including all plants from tropical, temperate and glacial regions.

Like Linnaeus, Adanson abstracted from plants the characteristics that defined degrees of identity and difference. Adanson's rejection of the binomial system resulted from the fact that Linnaeus' studies mainly considered the floral parts when classifying plants.

Despite being very rigorous, Adanson did not succeed in replacing the artificial system proposed by Linnaeus, a task that was left to the young apprentice Antoine Laurent de Jussieu (1748-1836). It was therefore with *Genera Plantarum* (1789) and other writings by Jussieu that the natural method took on greater proportions in botany in the mid-19th century.

Theorie Elemaintare de la Botanique, from 1813, by the Swiss Augustin Pyrame de Candolle (1778-1841), had also made a significant contribution to botanical nomenclature, where, in the first part of the work, De Candolle coined the term taxionomia , from the Greek *táxis:* "ordering"; *nómos:* "rule or law"; *ía:* "quality or property", thus meaning ordering rule.

In his work, De Candolle divided plants into two main groups: avascular plants and vascular plants. For De Candolle, the classification of plants should be based on anatomy and not physiology, as proposed by some scholars of the time.

In his book *Prodromus Systematis Naturalis Regni Vegetalis,* published in 1824, De Candolle described in detail his rules for naming plants, taking into account aspects

[5] Available at: http://gallica.bnf.fr/ark:/12148/bpt6k96488m> Accessed on: 19 Aug. 2015

of evolution, geography, agronomy and biometry. In this work, De Candolle classified and described all the botanical groups that existed at the time, which resulted in a large number of volumes, requiring his son Alphonse Louis Pierre Pyrame de Candolle (1806-1893) to continue his work, which was finalised in 1873.

The work *Lois de Nomenclature Botanique,* presented by Alphonse de Candolle in 1867, as well as Lineu's various contributions to the system of naming living beings, were standardised at the 1st International Botanical Congress, held in Paris in 1867. At that congress, an authentic code of botanical nomenclature was created, called the Paris Code. Since then, many other codes have been published, the latest in 2000 being the Saint Louis Code, which contains details of botanical taxonomy and nomenclature (RODRIGUES, 2009).

Likewise, the International Congress of Zoology in 1898 set up a commission to produce an International Code of Zoological Nomenclature, which was also based on the studies and system drawn up by Linnaeus. The first International Code of Zoological Nomenclature came into use in 1901.

From the works of many scholars, year after year, rules were created and altered for the classification and naming of living beings. The nomenclatural rules of post-Linean botany can be seen in the work of Rodrigues (2009), in which data on nomenclatural changes and their respective dates (from 1756 to 2001) were compiled.

The changes made at subsequent congresses arose from new studies and proposals to facilitate the system of identifying and classifying living beings. This new knowledge accompanied the methods and theories of the time, and many of the scholars were already pointing out the need for a more appropriate system for classifying living beings.

In the search for a more appropriate system for classifying living beings, German entomologist Willi Henning (1913-1976) was the first to propose a method for establishing kinship relationships between beings, based on Darwin and Wallace's theory of evolution (SANTOS, 2008). Claiming that organisms were genealogically related to each other from common ancestors, Henning gave rise to a method of classification that became known as phylogenetic systematics.

1.3 PHILOGENETICS

The major impetus for the emergence of phylogenetics was the acceptance and incorporation of Charles Darwin's evolutionary theories into the traditional classification system of living beings. Classification now took into account characteristics that reflected the evolutionary kinship between species.

Phylogenetics therefore uses phylogenies or phylogenetic trees, graphic structures that serve to represent the degree of evolutionary kinship between species or groups of living beings. Ernest Haekel (1834-1919) pioneered the use of this type of representation (RIDLEY, 2009), but phylogenetic systematics only emerged in 1950 with the publication by Willi Hennig of the work entitled "Elements of a Theory of Phylogenetic Systematics" (BRZOZOWSKI, 2014; RIDLEY, 2009).Phylogenetic systematics began based on the anatomical study of species, but after the emergence of molecular techniques it was possible to complement this area of science with information such as compounds produced by organisms, genetic material and more detailed studies of tissues (RIDLEY, 2009).

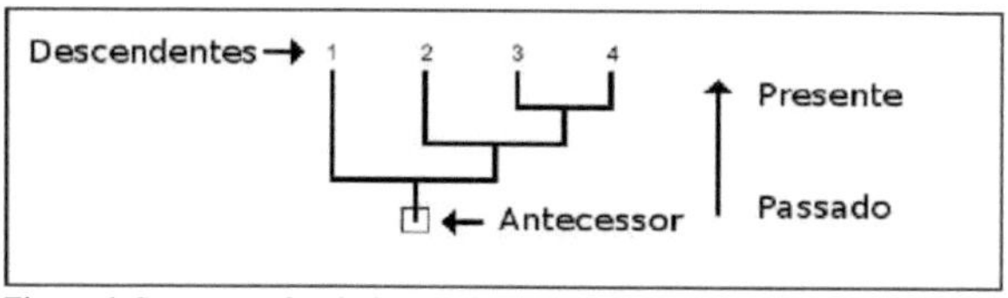

Figure 6: Structure of a phylogenetic tree Source: Understanding Evolution[6]

The structure of a phylogenetic tree starts from a trunk common to all beings and divides until it reaches isolated branches, which represent groups of living beings or species. The relationship with time is expressed from the common trunk, which represents the oldest lineage, and the higher up the tree the more recent the lineage represented (Figure 6).

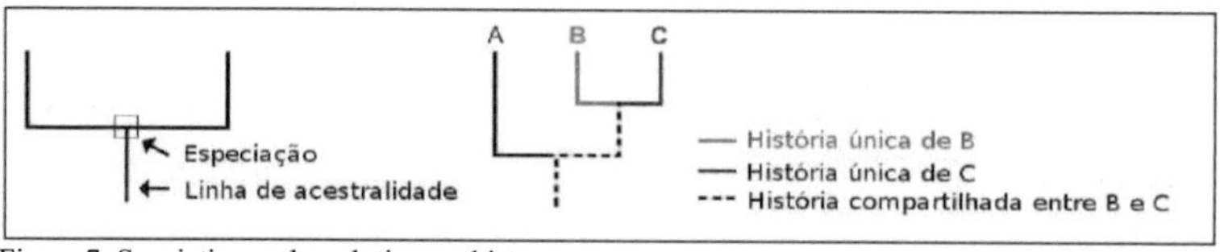

Figure 7: Speciation and evolutionary history
Source: Understanding Evolution[6]

[6] Available at: <http://evolution.berkeley.edu/evolibrary/article/evo_05> Accessed on: 20 Aug. 2015

In a phylogenetic tree, the division of a branch into two means that a species split into two new species at a certain time (speciation occurred - Figure 7). From the top to the bottom of the tree, it is possible to identify nodes (joins between the branches of the tree), which associate two beings with common ancestry. The lines common to two or more species represent the shared evolutionary history between these species (Figure 7).

The distinguishing feature of modern biological classification is that it takes into account the evolutionary similarities between species, often diverging from traditional classification, which was based exclusively on the study of the morphology of beings (FREEMAN, 2009). Because of this, there is a tendency to reorganise the traditional taxonomic categories (species, genus, tribe, family, order, phylum and kingdom), which are now called ciada, a term derived from cladistics. The term clade is used to designate a group of species derived from a common ancestor.

Cladistics is a science derived from phylogenetics and is therefore also based on the comparative anatomical study of beings, however, the structure used to represent kinship between beings is called a cladogram (FREEMAN, 2009).

Cladograms and phylogenetic trees are very similar; cladograms are constructed according to the principles of cladistics. One of these principles is that every species arises from the division of an ancestral species (RIDLEY, 2009).

A cladogram shows information about kinship over a period of time between three or more groups, but the names of the ancestors are never explicitly displayed (BRZOZOWSKI, 2012). So, as with trees

phylogenetic, each terminal branch of the cladograms must necessarily be a species or group of monophyletic beings (BRZOZOWSKI, 2014).

The representation of a phylogeny hardly encompasses the evolutionary relationships between all beings, being only a partial representation in which certain branches or nodes can be excluded, thus making it necessary to form subgroups, called monophyletic, paraphyletic and polyphyletic.

Monophyletic, paraphyletic and polyphyletic are types of groupings of living beings that have genealogical relationships with each other (Figure 8). Monophyletic groups are formed by just one parent entity and all its descendants (pink box in Figure 8); paraphyletic groups are formed by one entity and some of its descendants (green box in Figure 8), but not all of them; and polyphyletic groups are groups in which the common ancestor of the group members is not part of the group (blue box in Figure 8) (BRZOZOWSKI, 2012). Numerous challenges have confronted researchers involved in the study of the classification of living beings. Among them is the choice of the relevant aspects for classification,

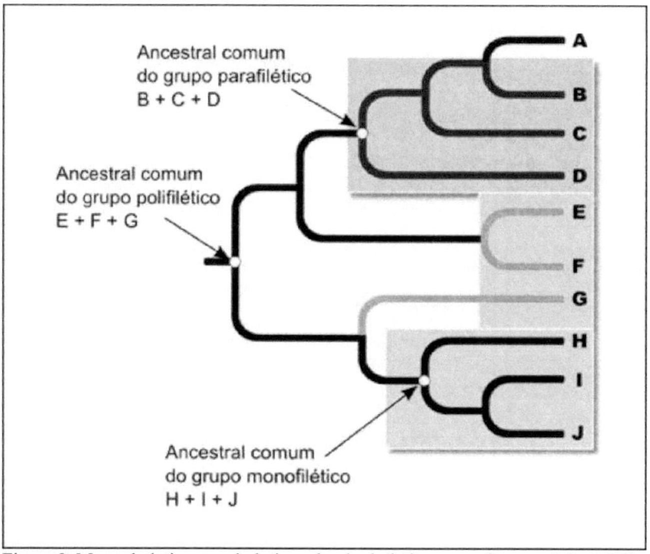

Figure 8: Monophyletic, paraphyletic and polyphyletic groups Source: Adapted from Sadava et *al.* (2009)

which takes into account the evolutionary characteristics of living beings, be they morphological, functional or molecular (SANTOS, 2012).

Terms such as homology and analogy have emerged in the context of modern systematics to designate relationships between organs in living beings. Homologous organs are therefore body structures that develop in a similar way in different species, but which come from a common ancestor. Despite having a similar embryonic origin, homologous organs can have different morphology and functions (RIDLEY, 2009).

Analogous organs are structures that have appeared independently in different groups of living beings, but have been modified to adapt to similar ways of life. Analogous organs are therefore different from an anatomical point of view and in terms of their embryonic origin, but they can have similar functions.

The study of anatomical structures and the determination of homologous and analogous organs has provided clues to the evolutionary origin of species and is widely used in phylogenetic and cladistic studies of living beings (FREEMAN, 2009).

The biological classification of species has kept pace with advances in the fields of molecular biology and computing, so that molecular comparisons between different groups of living beings are now possible.

Despite the widespread use of morphological characteristics to build cladograms, representations based on molecular analyses of proteins and nucleic acids are increasingly being used. Genes are the main targets of new forms of classification. Classifications based on the study of DNA will be discussed in more detail in Chapter 3.

CHAPTER 2

GENOME-BASED CLASSIFICATIONS

Since the earliest forms of classifying living beings, many transformations have taken place, with classification methods no longer being based solely on the morphological characteristics of living beings, but taking into account chemical and cytogenetic characteristics.

Advances in genetics, such as the development of DNA sequencing techniques, the creation of databases with genetic information on various living beings and computer tools for processing genetic data, have made it possible to classify and identify living beings in new ways. These new forms of classification, especially those based on molecular analyses, have produced a number of conflicts in comparison with older forms of classification, which focused solely on the morphology of living beings (FREEMAN, 2009).

In recent years, taxonomy has been strengthened by analysing genetic codes, in the hope that the genetic load of species can provide important information about their degree of kinship.

For phylogenetic analysis, the process of analysing the genetic code in order to demonstrate an association between species, several hundred genes have been combined in matrices in an attempt to establish increasingly precise relationships between beings from the major taxonomic groups. This mechanism of phylogenetic analysis has been applied to different groups of living beings: animals (ROKAS *et al.,* 2005), plants (PHILIPPE *et al.,* 2005) as well as various studies on fungi and bacteria.

The study of the phylogenetic relationships of beings has used a variety of techniques, from those involving molecular biology, such as the amplification of nucleotide chains, to those involving mathematical analyses associated with various *software programs* for constructing phylogenetic trees (TATENO, 1994). Determining which genes are suitable for analysing kinship between species has also been a point of fundamental importance and intense debate in scientific circles.

According to Russo (2012), in order to determine phylogenetic relationships, several factors must be taken into account, such as whether the analysis will be of nucleotide or amino acid sequences, checking the rate of gene evolution and the divergence time between species. Different data sets chosen and even different reconstruction methods can result in different phylogenies.

Various genes have been exhaustively tested in order to determine their potential to provide genetic comparisons and identify species. It should be noted that different genes are used for different groups of organisms. In plants, for example, *rbcL* is widely used (MACHADO, 2012), and even within a larger taxonomic group, subgroups can use different genes to infer phylogenetic relationships, as in the *Myrtaceae* family, for which the gene most suited to determining kinship between family members is *matK* (MACHADO, 2012).

Choosing the right genes to build phylogenies requires investigations that take homology into account, which, in molecular terms, refers to the fact that two genes descend from a common ancestor (MACHADO, 2012). Determining this homology between genes is essential, but not sufficient to determine a phylogeny.

The amino acid sequences that make up genes are influenced by evolutionary mechanisms, leading to small changes in the genetic code. These changes occur at different rates between groups of living beings and can lead to nucleotide insertion or deletion events. The relationship between time and divergence between the sequences that make up a gene is called the rate of evolution.

The rates of gene evolution also give important clues about the choice of appropriate genes for phylogeny; cytochrome C, for example, is among the most conserved genes, less variable between organisms and therefore may not provide relevant information for assembling phylogenies of some groups of living beings (RUSSO, 2001). Various computer algorithms make it possible to carry out multiple alignments between nucleotide or amino acid sequences, an activity that makes it possible to establish the rate of gene evolution, as well as inferring a degree of similarity between sequences.

Alignment between sequences is the activity of comparing all the characters in a sequence A against the characters in a sequence B. Algorithms can analyse all the characters between sequences. They usually generate an alignment score, which is made up of rewards (if the characters of the sequences in the same position match) and punishments (if the characters of the sequences in the same position don't match, or if adjustments need to be made to the analysed sequences). The scoring mechanism between the alignment of two DNA sequences can be seen in figure 9.

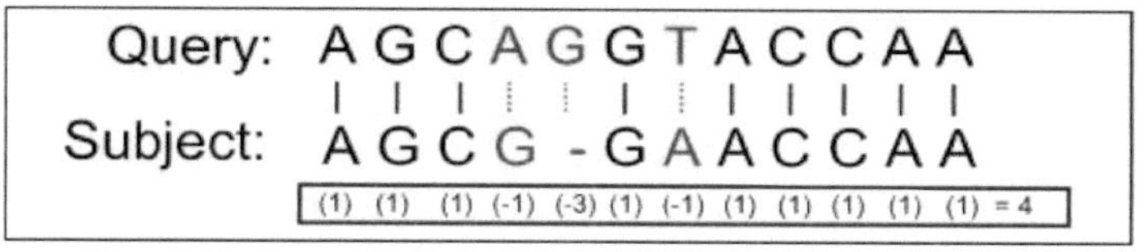

Figure 9: Scoring mechanism in a DNA sequence alignment.

In figure 9, the characters after the word *Query* are part of *DNA* sequence A and the characters after the word *Subject* are part of DNA sequence B. Every time a character in sequence A is the same as a character in the same position in sequence B, the score is increased (in the example in figure 9, it is 1), whenever a character in sequence A diverges from a character in sequence B, a negative value is assigned (-3) and when there is a space instead of a character to be compared, a negative score is also assigned, but less (-1) than in the case of divergences between characters.

The gaps in the sequences may be due to a missing nucleotide in the original DNA sample, or due to the alignment algorithm's need to insert them for a better fit between the sequences. Both the spaces inserted by the algorithm and those from the original sequence are quantified and contribute to the final alignment score. The example shown in figure 9 occurs with an alignment between two sequences, however, to align multiple sequences (more than two) the algorithms use matrices that perform various calculations to determine scores and degrees of similarity.

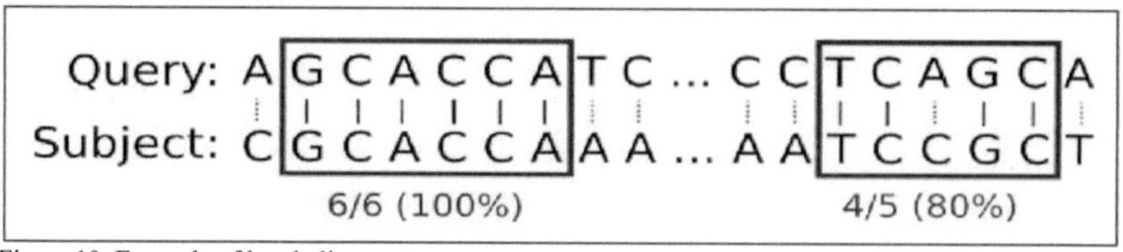

Figure 10: Example of local alignment

The alignment process can be done in two ways: local alignment or global alignment. In local alignment, sections of the sequences are superimposed in search of some similarity, resulting in a set of similar subsequences. In simplified terms, the local alignment of two sequences considers arbitrary segments, with no penalties for misaligned parts (SMITH & WATERMAN, 1981). Global alignment considers the entire string, resulting in a similarity index between the two entire sequences. The process of local alignment and global alignment are exemplified in Figure 10 and 11, respectively.

Among the most popular sequence alignment software are *BLAST (Basic Local Alignment Search Tool)* (ALTSCHUL, 1990) - maintained by the *NCBI*, *Clustal* (HIGGINS, 1988) and MUSCLE *(MUltiple Sequence Comparison by Log-Expectatiori)* (EDGAR, 2004) both maintained by the European Molecular Biology Laboratory.

```
  Query: A G C A C G A T C C G C A T C A G C A
         : | | | | : | : | : | : | | | | | | :
Subject: C G C A C A A A C G G A A T C C G C T
                     13/19 (68.42%)
```

Figure 11: Example of global alignment

While local alignment has greater relevance in the process of species identification, the global alignment of multiple sequences can help determine the rate of evolution and variation of genes in a specific group of organisms.

To better exemplify how to find clues about the kinship relationships between species, some nucleotide sequences were obtained from *NCBI Genbank and* used at various times in this work for alignment and phylogeny reconstruction purposes. These sequences correspond to the *rbcL* gene of *Zea mays (emb\Z11973.1\)*, *Nicotiana tabacum (emb\Z00044.2\)*, *Pinus thunbergii (dbj\D17510.1\PINCPTRPG)* from the *Plantae* kingdom *and Thiobacilus denitrificans (gb\CP000116.1\)* a bacterium, obtained by consulting the *NCBI Genbank* database.

```
CLUSTAL 2.1 multiple sequence alignment


Zea_mays                       ATGTCACCACAAACAGAAACTAAAGCAAGTGTTGGATTTAAAGCTGGTGT 50
Nicotiana_tabacum              ATGTCACCACAAACAGAGACTAAAGCAAGTGTTGGATTCAAAGCTGGTGT 50
Pinus_thunbergii               ATGTCACCAAAAACAGAGACTAAAGCTAGTGTCGGATTCAAAGCTGGTGT 50
Thiobacillus_denitrificans     ATG--GCTGT---CAAAACCTATAGC----------------GCCGGTGT 29
                               ***   *      ** *  *** ***                ** *****

Zea_mays                       TAAGGATTATAAATTGACTTACTACACCCCGGAGTACGAAACCAA-GGAT 99
Nicotiana_tabacum              TAAAGAGTACAAATTGACTTATTATACTCCTGAGTACCAAACCAA-GGAT 99
Pinus_thunbergii               TAAAGATTACAGATTAACTTATTATACTCCTGAATATCAGACCAA-AGAT 99
Thiobacillus_denitrificans     GAAGGAATACCGGCAGACTTACTGGATGCCCGAATAC-ACGCCGCTGGAT 78
                                ** ** **       ***** *  *  ** ** **  *  **    ***

Zea_mays                       ACTGATATCTTGGCAGCATTCCGAG-TAACTCCTCAGCTCGGGGTTCCGC 148
Nicotiana_tabacum              ACTGATATATTGGCAGCATTCCGAG-TAACTCCTCAACCTGGAGTTCCAC 148
Pinus_thunbergii               ACAGATATCTTGGCAGCATTCCGAG-TAACTCCTCAACCCGGGGTGCCAC 148
Thiobacillus_denitrificans     ACCGACATCCTGGC-GTGCTTCAAGATCACCCCGCAGGCGGGCGTTGACC 127
                               ** ** **  **** *    * * ** * ** ** **    ** **    *

Zea_mays                       CTGAAGAAGCAGGAGCTGCAGTAGCTGCGGAATCTTCTACTGGTACATGG 198
Nicotiana_tabacum              CTGAAGAAGCAGGGGCCGCGGTAGCTGCCGAATCTTCTACTGGTACATGG 198
Pinus_thunbergii               CTGAAGAAGCGGGAGCAGCAGTAGCTGCTGAATCTTCCACCGGTACATGG 198
Thiobacillus_denitrificans     GTGAAGAAGCAGCCGCCGCCGTGGCCGCCGAGTCGTCGACCGGCACCTGG 177
                                ********* *  ** ** ** ** ** ** ** ** ** ** ** ***

Zea_mays                       ACAACTGTTTGGACTGATGGACTTACCAGTCTTGATCGTTACAAAGGACG 248
Nicotiana_tabacum              ACAACTGTATGGACCGATGGACTTACCAGCCTTGATCGTTACAAAGGGCG 248
Pinus_thunbergii               ACCACTGTTTGGACCGATGGACTTACTAGTCTCGATCGTTACAAGGGGCG 248
Thiobacillus_denitrificans     ACCACCGTGTGGACCGACCTGCTGACCGACCTCGACTACTACAAGGGCCG 227
                               ** ** ** ***** **    ** **    ** **    ***** ** **
```

Figure 12: Part of the result of the multiple alignment of sequences in *CLUSTALW*.

Figure 12 shows the result of the multiple alignment of *Zea mays, Nicotiana tabacum, Pinus thunbergii* and *Thiobacilus denitrificans*, which consists of the name of the organism followed by the aligned section of the sequence, the number at the end of each line indicating the number of characters aligned up to that line of the result. The characters represent spaces in the sequence and the asterisks ("*") displayed mark the positions where all the characters of the sequences involved in the sequence are located.

alignment were identical.

In the aligned sections of the sequences in figure 12, it is possible to see a large divergence between the sequence of *Thiobacilus denitrificans* and the other organisms.

There was little divergence between the organisms that make up the *plantae* kingdom in terms of the characters that represent the gene. Table 1 corroborates the result previously analysed in figure 12: the scores of *Thiobacilus denitrificans* aligned with the other organisms were considerably lower than the alignment of sequences of organisms from the same kingdom.

The alignment results in Figure 12 and Table 1 allow us to preliminarily establish the idea that the *rbcL* gene is a favourable candidate for distinguishing species at the kingdom level, since the alignment score between the sequences of beings from the *plantae* kingdom was considerably higher than the score obtained by the bacterial sequence. It is possible to establish that the organism *Thiobacilus denitrificans* is not part of the plant kingdom, although using more accurate statistical techniques it is possible to infer evolutionary relationships between *Thiobacilus denitrificans* and the other organisms analysed.

Table 1: Score of the multiple sequence alignment partially represented in figure 12.

SeqA	In the	Length	SeqB	Name	Length	Score
1	Th iobac illu s_den itrifican s	1422	2	Zea_mays	1431	65.19
1	Th bbac illu s_den itrif ic an E.	1422	3	Pinirs_thunbergii	1426	66.39
1	Th iobac illus_der itrifican s	1422	4	N icotianajabac u m	1434	64.63
2	Zea_mays	1431	3	Pinusjhunbergii	1428	86.06
2	Zea mays	1431	1	N icatianajabac u m	1434	86.44
3	Pinus_thunbergii	1426	4	Nicotianajabacun'	1434	86.9

Local alignment can be applied between nucleotide or nucleic acid sequences to identify living beings at the species level. Cloud *et al.* (2002) state that the main strategies for diagnosing agents that cause hospital-acquired infections by sequencing consist of analysing the regions

encoding *16S* ribosomal *RNA (rDNA).* The technique offers the possibility of rapid bacterial identification by targeting the conserved *16S rDNA* region. Some studies (MAGNANI, 2010; RUPPENTHAL *et al.,* 2005) point to the use, through sequencing and alignment, of the *16S rRNA* gene against databases for the diagnosis of bacteria.

Identifying bacteria, especially pathogenic ones, is possible by locally aligning a sequence against a sequence database. Public databases containing genomic information on species can be accessed free of charge by anyone, such as *NCBI GenBank* - maintained by the *National Centre for Biotechnology Information* and *UniProt* - maintained by the *Protein Information Resource (PIR).* As well as providing hundreds of gigabytes of data from various genomes and information on living beings, they also provide tools for carrying out sequence alignments.

The purpose of aligning sequences can be both to identify living beings and to relate species evolutionarily. For the study of evolution, the techniques used are different from those that make it possible to identify beings.

With the evolutionary aspect in mind, some tools have been developed to facilitate phylogenetic analyses based on protein or *DNA* sequences. In this sense, *MEGA - Molecular Evolutionary Genetics Analysis* (KUMAR *et al.,* 2004) is one of the most widely used pieces of software for carrying out phylogenetic analyses.

MEGA has a set of tools capable of carrying out statistical analyses, enabling sequence alignments, reconstruction and visualisation of phylogenetic trees, tests on lists of evolutionary hypotheses, estimation of divergences between sequences, acquisition of sequences based on web services, as well as expert systems that generate natural language descriptions of the analysis methods chosen by the user (TAMURA *et al.,* 2011).

The use of *MEGA* eliminates the need for complementary programmes to build phylogenetic trees, since it is integrated with the main algorithms for aligning protein and nucleotide sequences. The software works on *Windows* and *MAC OS* platforms, is free software and can be used by anyone with basic computer skills.

It is therefore possible to create phylogenies easily with *MEGA,* based on relevant *DNA* sequences from a wide variety of organisms. However, in order to use the programme, it is necessary to obtain the *DNA* or protein sequences in a file that *MEGA* recognises (the *FASTA* format is commonly used, figure 13).

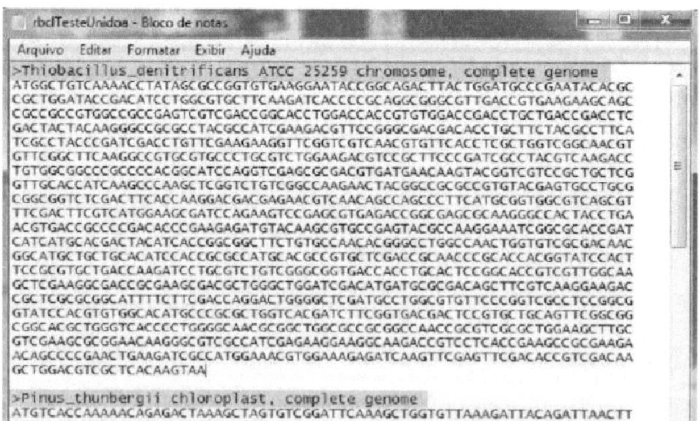

Figure 13: File in FASTA format containing various DNA sequences. The highlighted area corresponds to the title related to the sequence.

MEGA has algorithms for translating DNA into protein or protein into DNA, allows you to consult *Genbank and* the *NCBI* website in an internal *browser, and* integrates *BLAST* and *MUSCLE,* automating sequence alignment.

In *MEGA,* it is also possible to choose between different methods for constructing phylogenetic trees. *Maximum likelihood* (FELSENSTEIN, 1981), *Neighbour- Joining* (SAITOU & NEI, 1987), *Minimum Evolution* (RZHETSKY; NEI, 1993), *UPGMA (Unweighted Pair Group Method with Arithmetic Meari)* (SNEATH & SOKAL; 1973) and *Maximum Parsimony* (ECK *et aL,* 1966) are the methods that can be used in *MEGA*.

be chosen. An in-depth discussion of the best methods for constructing phylogenetic trees can be seen in Tateno (1994).

Figure 14 shows the phylogenetic tree generated using the *Maximum Likelihood* method in the *Tree Explorer* window of the *MEGA* programme with the sequence data used in the alignment in figure 12. Trees with data from the same sequences using different phylogenetic reconstruction methodologies, as well as considerations on using the *software*, can be seen in Appendix B.

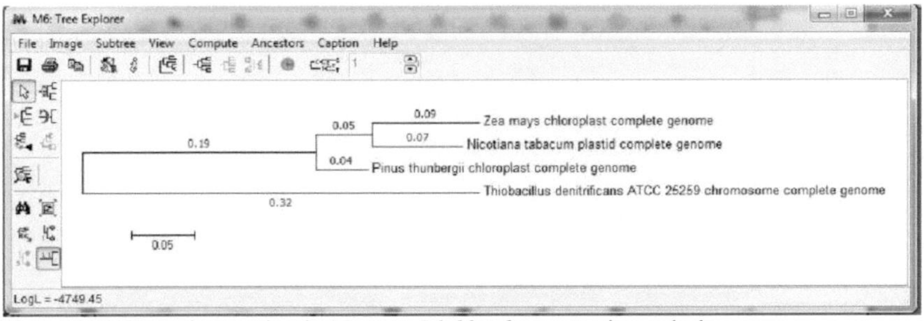

Figure 14: Phylogenetic tree using the *Maximum Likelihood* reconstruction method.

Similar to *MEGA, PAML - Phylogenetic Analysis by Maximum Likelihood* (YANG, 1997) is another tool that also deserves mention because it also has a set of programmes capable of generating phylogenetic trees. *PAML* is free *software* and open source, which allows any developer to collaborate on its improvement. It works on *Windows, MAC OS* and *Linux* platforms.

The emergence and evolution of computer tools is making it possible to make considerable changes to the classification of living beings. This change in the classification of living beings can already be seen in various works and textbooks.

An example of these changes, the result of phylogenetic classifications, was proposed by the *APG (The Angiosperm Phylogeny Group)* in *"An ordinal classification for the*

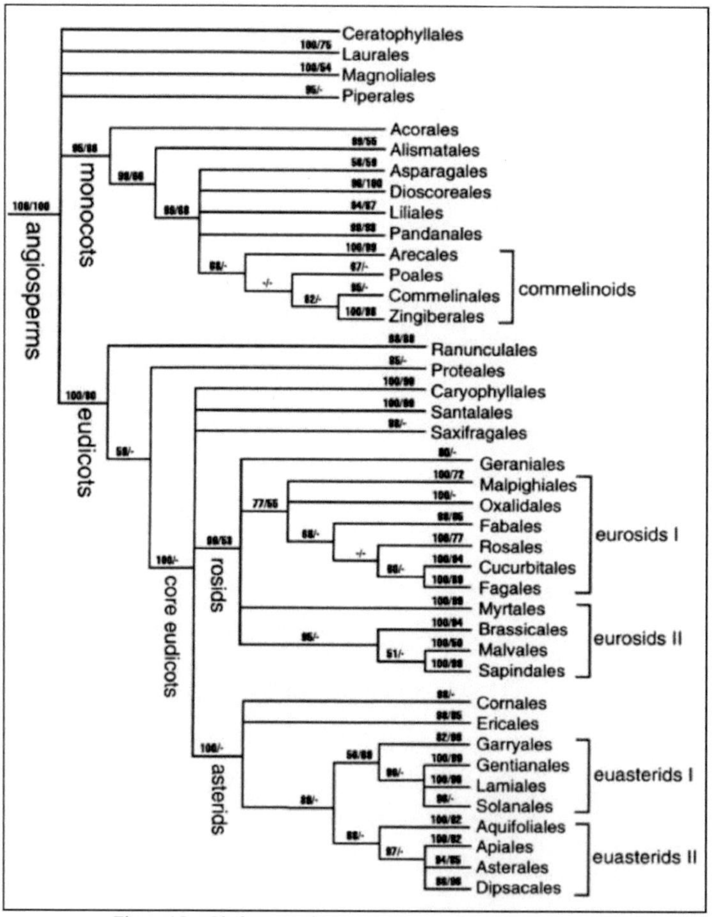

Figura 15: Phylogeny of orders and some families in the *APG* Source: Adapted from BREMER *et al.*, 1998.

families offiowering plants" in 1998. In this first work, the group presented a classification system initially composed of 462 families organised into 40 orders, supposedly monophyletic (BREMER, 1998). In their studies, the *APG* group used a compilation of literature containing molecular analyses of *rbcL, 18S rDNA, atpB and matK* genes, among others, *to* carry out a phylogenetic classification.

Following the publication of *"An ordinal classification for the families of flowering plants"*, the APG carried out two updates: *"An update of the Angiosperm Phylogeny Group classification for the orders and families of flowering plants: APG II"* (BREMER, 2003) and *"An update of the Angiosperm Phylogeny Group classification for the orders and families of flowering plants: APG III"* (BREMER, 2009). Figure 15 shows the phylogenetic relationships established between the angiosperms in the first *APG* publication.

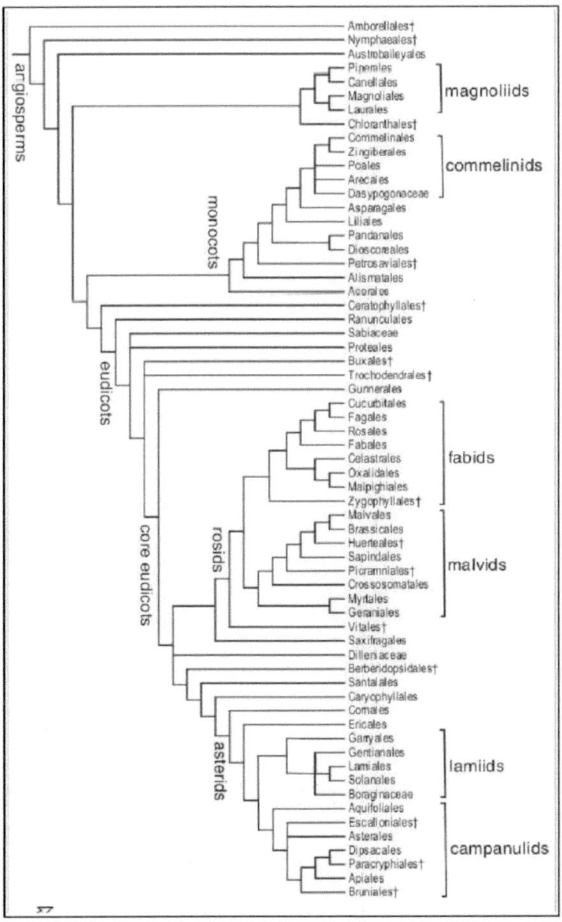

Figure 16: Phylogeny of orders and some families in *APG III* Source: Adapted from BREMER *et aL*, 2009.

Considerable changes can be seen between the results of the first study in 1998 and the latest (figure 16) in 2009. Among these changes is the appearance of new orders, since the first paper categorises 40 orders and the latest update lists 63 orders. The phylogeny shown in figure 16 refers to the work published by Bremer *et al.* in 2009, and the orders marked with the symbol (t) are newly recognised by the *APG.* In addition to the appearance of new classes, some new cites have been established, such as

for example, which associates the classes *Oxalidales* and *Malpighiales* with a common ancestor, a question not yet fully established in the first publication.

The *APG* group states that it is almost impossible to reach universal agreement on the changes involving species and that, due to the large amount of work in the area of phylogeny, the proposed changes to the *APG* classification are modest (BREAMER, 2003).

The use of genetic analyses in the construction of phylogenetic relationships has generated changes and, as a result, discussions among the scientific community. Phylogenetic analyses can be seen in various studies, including Zhang *et al.* (2001) who carried out a phylogenetic and chemotaxonomic study of the *Thermomonosporaceae* family. In their work, Zhang *et al.* (2001) concluded, after sequencing and aligning the genetic codes based on *16S rDNA, 16S-23S rRNA* and *23S rDNA* sequences, that four of the species studied had been transferred from the genus *Actinomadura* to the genus *Actinocorallia.*

> Based on both phylogenetic and chemotaxonomic evidence, we propose the transfer of four *Actinomadura* species to the genus *Actinocorallia: Actinomadura libanotica* as *Actinocorallia libanotica* comb. nov., *Actinomadura aurantiacs Actinocorallia aurantiaca* comb. Nov., *Actinomadura glomerata* as *Actinocorallia glomerata* comb. nov. and *Actinomadura longicatena* as *Actinocorallia longicatena* comb. Nov. (ZHANG, 2001, p. 380)

The changes, especially to the genera, have a direct impact on the name of the beings, since the genus is part of the scientific name of the species.

2.1 DNABARCODE

In 2003, Paul Herbert, a researcher linked to the Zoology department at the University of Guelph in Canada, in an article resulting from pioneering studies, announced the possibility of distinguishing animals through small pieces of *DNA*.

Herbert *et al.* (2003) called the short *DNA* sequences used to identify species *"barcodes"* because this stretch of *DNA* embedded inside each cell of the organism can actually function, by analogy, as a kind of barcode for living beings, which can be unique to each species.

Barcodes are known as market product identifiers and generally use 10 numbers alternating in 11 positions, resulting in 100 billion unique identifiers. The same proposal can be used to identify living beings, but using the four chemical bases (Adenine, Guanine, Cytosine and Thymine) that make up *DNA* sequences. If only 15 nucleotides are considered, more than 1 billion unique identifiers could be formed, which would be enough to discriminate between all known living beings.

The idea behind *barcoding* is to make it easier to identify living beings without the need for a whole specimen (HERBERT, 2003). This would allow scientists in the field to quickly identify beings using handheld sequencers without having to resort to books, identification keys or professional taxonomists.

One of the first difficulties with *barcoding* was determining an ideal gene that should fulfil three criteria: (I) have significant genetic variability between species, (II) be short in length to facilitate *DNA* extraction and amplification, and (III) have conserved regions for the development of universal primers (KRESS *et al.,* 2005).

Mitochondrial ribosomal coding genes such as *rDNA 12S* and *rDNA 16S* have been evaluated, but their use in taxonomic analyses is restricted due to the prevalence of insertions and deletions, which generally make sequence alignment difficult (DOYLE & GAUNT, 2000).

Some characteristics suggest that the *COI* gene *(cytochrome c oxidase* I) could be the ideal target to act as a *barcode* for animal species. COI has advantages such as: (I) a lack of *introns* (non-coding regions), an inherent characteristic of mitochondrial genome sequences (SACCONE *et al.,* 1999), (II) it has robust *primers*, which allows specific segments of the DNA sequence to be recovered (FOLMER *et al.,* 1994) and (III) changes in the amino acid sequence of this gene occur more slowly than any other mitochondrial gene (LYNCH & JARRELL, 1993).

In their work, Hebert *et al.* (2003) created a COI gene profile for seven different phyla of the animal kingdom, based on the analysis of 100 representative species. They also analysed the class Hexapoda (as it has a large number of representatives in the planet's biodiversity), some insect orders with a focus on lepidopterans, the latter group with the aim of correctly identifying the species (the group was chosen because it has little divergent DNA sequences between the families of this order).

With the results of the tests, Herbert *et al.* (2003) demonstrated that the amino acid sequences in *COI* are sufficiently reliable for the identification of higher taxonomic categories, since the beings analysed were grouped in the correct order or phylum. When it came to identifying lepidopterans, the tests showed 100% success for species-level identification. Despite the small sample analysed, considering the biodiversity of the planet, it was possible to establish that the *cytochrome c oxidase I (COI)* gene can serve as the core of a global system for animal identification (HEBERT 2003, p 313).

Identification based on the *COI* gene will undoubtedly provide taxonomic solutions that exceed those that can be obtained through morphological studies (HERBERT *et al.,* 2003). In their work, Herbert *et al.* (2003) also propose the creation of a specific database for the COI gene. The conclusions of Herbert *et al.* (2003) can be reinforced in several other studies published more recently (RESCH *et al.,* 2014; JO *et al.,* 2014; SAITOH *et al.,* 2015).

After the publication that a gene had been found to discriminate between animal species (HERBERT *et al.,* 2003), the efforts of many researchers turned to determining a gene that could distinguish and classify plant species. It was essential that a barcode was created to discriminate between plants, given that this kingdom has a great variety of species, some of which show minimal differences, as well as the impossibility of identification in situations where the plant does not have the vegetative elements necessary to recognise it.

Kress, one of the pioneers in research into a gene to identify plants, was aware that the COI gene was not suitable for discriminating between plants at the species level, because mitochondrial genes in plants change more slowly than in animals (BERGTHORSSON, 2003), so a different gene was needed.

The experts soon turned their efforts to the genome not found in animals, then began analysing the chloroplasts of plant species in the hope that they could provide an ideal target gene for *DNA barcoding* in plants.

In molecular research on plants at the species level, the ribosomal region *ITS (Internal Transcribed Spacer)* was most commonly sequenced. In addition, because it is used in many phylogenetic studies (KRESS *et al.,* 2005), there was already a large amount of sequence data from this region available in public databases such as Genbank, which at the time had more than 36,000 angiosperm sequences.

According to a study carried out by Kress *et al.* in 2005, the *ITS* spacer was the most suitable candidate for *DNA barcoding*. Among the advantages were: (I) the fact that it can be amplified in two small contiguous fragments *(ITS1* and *ITS2)*, a locus called *5.8S* - this is useful for degraded samples (which have suffered damage after contact with sunlight, microorganisms or solvents) and (II) although it has phylogenetic signals for discrimination at the level of orders and phyla, the 5.8S locus can be useful for discovering levels of kinship between species through more specific alignments (KRESS *et al.,* 2005).

In addition to the *ITS* (locus *5.8S locus),* Kress and colleagues analysed the *rbcL* gene *and* seven other plant *DNA* intergenic spacer regions:, *trnK-rpsl6, trnH-psbA, rpl36-rps8, atpB-rbcL, ycf6-psbM, trnV-atpE, trnC-ycf6, psbM- trnD, and tmL-E* The studies were carried out on specimens from Plummers Island in the USA, and the focus was on determining the best target for *DNA barcoding* in plant species.

Among the regions analysed, *rbcL* showed the lowest divergence (0.83%), but the length of the sequence (generally 1428 base pairs) and the amplification success rate (95%) were not the best. On the other hand, the *trnH-psbA* region was 100% successful in amplifying the sequences, showed the highest divergence value in most of the genera and species analysed and the region is made up of a smaller number of base pairs.

Kress and his collaborators mainly considered the highest divergence value between the sequences to be the most important criterion in distinguishing living beings at species level, since 6 of the 8 genera and 11 of the 14 species analysed in their work could be identified on the basis of the *trnH- psbA* analysis.

The high frequency of nucleotide differences in *trnH-psbA* (SHAW *et al.,* 2005), coupled with the short length of the sequences, provide a significant advantage for using the region in *DNA barcoding.* However, the *trnH-psbA* region is of limited use for phylogenetic reconstruction due to the short length of the sequence and the high number of insertions or deletions (*indels)* (ACKERFIELD & WEN, 2003; YAMASHIRO *et al.,* 2004).

In their considerations, Kress *et al.* (2005) believe that *ITS* and *trnH-psbA* serve as good starting points for large-scale testing of *DNA barcoding* of plant species, through a large sample of angiosperms. He also suggests that a good test would be to expand taxon sampling by applying both *ITS* (locus 5.88) and *trnH-psbA* for *DNA barcoding* in Costa Rican flowering plant species.

In the tests carried out by Kress and his collaborators, it became clear that the sequences of the 5.8S locus of mosses, liverworts and other distant relatives of flowering plants showed a lot of variation. Species such as ginger and tomato showed very different sequences (PENNISI, 2007).

With the functional limitations of the 5.8S *ITS* locus for recognising plants, additional genes present in plant plastids have been evaluated in order to complement the *DNA barcode.* Among the most commonly sequenced in plant species, for systematic and phylogenetic purposes, were: *rbcL,* intergenic spacer *trnL-F, matK, ndhF* and *atpB* (KRESS *et al.,* 2005).

Although all the genes mentioned above have a high degree of success in determining evolutionary relationships between species, the *rbcL* gene had already been suggested by Blaxter as a candidate for a *DNA barcode* for plants (BLAXTER, 2004).

In 2009, Kress and other researchers, now part of a group called CBOL

(Consortium for the Barcode of Life) Plant Working (GROUP, 2009), which was created in 2004 with the aim of building a *DNA barcode* library for all eukaryotic life (RATNASINGHAM & HEBERT, 2007), carried out a broader study on the best target for *DNA barcoding* in plants. The focus of the work was to analyse the main regions for use in *DNA barcoding, and* CBOL then went on to analyse the genes that were the subject of discussion among the scientific community.

The regions evaluated were four coding genes *(matK, rbcl, rpoB, rpoCl)* and three non-coding spacers *(ATPF-atpH, trnH-psbA,* and *psbK-psbl).* To do this, the researchers used data from 907 samples, representing 445 angiosperms, 38 gymnosperms and 67 cryptogams, following some *CBOL* standardisation criteria (amplification success, length and sequence divergence).

The results of the tests carried out showed that in six of the seven loci analysed in angiosperms, the success rate for *PCR* amplification and sequencing was between 90% and 98%, the lowest success rate was for the *psbK-psbL* region, which achieved 77% success. In terms of sequence quality, the *rbcL, rpoCl* and *rpoB* loci were the best, while the *tmH-psbA* and *psbK-psbl* intergenic spacers performed the worst. Combining the seven loci analysed, 73% of the sequences were successfully discriminated, combinations with two loci successfully discriminated between 59% and 75% of the sequenced samples, while combinations of three loci discriminated between 65% and 76%.

Among the loci analysed based on the *CBOL* criteria, four of the seven loci were discarded as *DNA barcode* targets. The *rpoCl* and *rpoB* loci performed well in terms of universality *(*robust *primers*) and sequence quality, but had low discriminatory power, *ATPF-atpH* was below average in terms of sequence quality and *psbK-psbl* showed a low success rate in terms of sequencing (HOLLINGSWORTH *et al,* 2009).

The remaining loci had desirable points for the *DNA barcode* system, but none of the three had all the necessary attributes, so it was necessary to combine two or even three loci (as suggested by some *CBOL* researchers*)*. In the results obtained by Hollingsworth *et al.,* 2009, the combination of three loci did not improve discrimination compared to the combination of two loci. The *rbcL* and *matK* genes combined were 72% successful in discriminating species.

> We therefore propose *rbcL+matK* as the standard *barcode* for land plants. This combination represents a pragmatic solution to a complex trade-off between universality, sequence quality, discrimination and cost (HOLLINGSWORTH *et al.,* 2009 p. 12796).

Discussions about the best regions for the *barcode* were intense.

Within the same research group, opinions and conclusions varied. While some, in independent studies, suggested that the *rbcL* gene was not suitable, others reinforced the results of Hollingsworth *et al.* (2009) regarding the choice of *matK.* The proposal for a *barcode* using three loci was also hotly debated.

In some cases, only one gene is needed to solve specific problems. The *trnL* gene, for example, for use as a *DNA barcode* has a relatively low resolution index, but when applied in specific contexts, such as species from a single ecosystem, or commonly consumed plants, it has a high degree of resolution (TABERLET *et al.,* 2007). However, genes that are commonly used to identify isolated groups may not be useful for taxonomists, who need the genes to provide more information about the degree of kinship between species.

The types of situations mentioned above culminate in a conflict of needs which does not always contribute to the evolution of *DNA barcode* research. In the search for an effective DNA barcode, *CBOL* relied on the collaboration of more than 200 organisations from 50 countries. Estimating that the volume of data for a *DNA bar*code library covering all species would reach more than 100 million records, it planned an enterprise-scale *software* infrastructure for handling and processing such a large volume of data.

According to Ratnasingham (2007) the key characteristics for the operation of a system that can provide identification via *DNA barcode* include the requirement for a persistent link between a *DNA barcode* sequence and its sample of origin in a secure environment that stores, organises and consults such records, accessible to the entire community. In addition, it was necessary to guarantee the quality and standardisation of the records through a robust computational platform.

Through dialogues with the main genomic data repositories such as

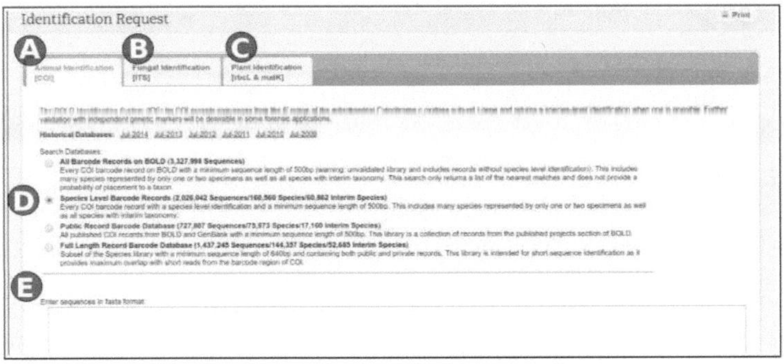

Figura 17: Details of the species identification page in BOLD.

NCBI and other organisations involved in the same line of research have established standards for the creation of the *Barcode of Life Data System (BOLD)* - a bioinformatics platform that supports all phases of the analytical journey from sample collection to *DNA barcoding.*

The *BOLD* platform currently has collaborators in various research centres in 138 countries and has 5,132,521 specimen records, of which 3,761,569 already have *DNA barcodes* (BOLD, 2015). In *BOLD* it is possible to use various tools to analyse genomic data, as well as the possibility of consulting species records.

On the *BOLD* identification page (Figure 17) you can choose between three tabs: animal identification using the *CO1* gene (A), identification of beings from the fungi kingdom using the *BLAST* algorithm with *ITS* sequences (B) and plant identification using the *rbcL* and *matK* genes (C). To identify animals, it is also possible to opt for the search database (D), which can also include the *Genbank* data repository. The marker (E) in figure 17 is where the *DNA* sequence is inserted.

The result of the query shown in figure 18 is the plastidial genome of

Figura 18: Result of the *BOLD* enquiry.

Nicotiana tabacum. (A) shows the type of query made, in this case to the *matK* and *rbcL* genes, followed by a graph (B) showing the highest alignment scores indicating the highest degrees of similarity. Marking (C) indicates the *e-value* graph where the result closest to zero is the best value. In (D) a list is displayed, ordered by similarity, showing the taxonomy of the plant, the score, similarity and *e-value,* as well as a status and a *hyperlink* leading to a page with a more detailed description of the organism. At the bottom of the page you can visualise the most significant alignments in detail.

The species record, present in the *hyperlink* at the end of each item in the results list (figure 18 - D), contains different categories of information that will be analysed in figure 20.

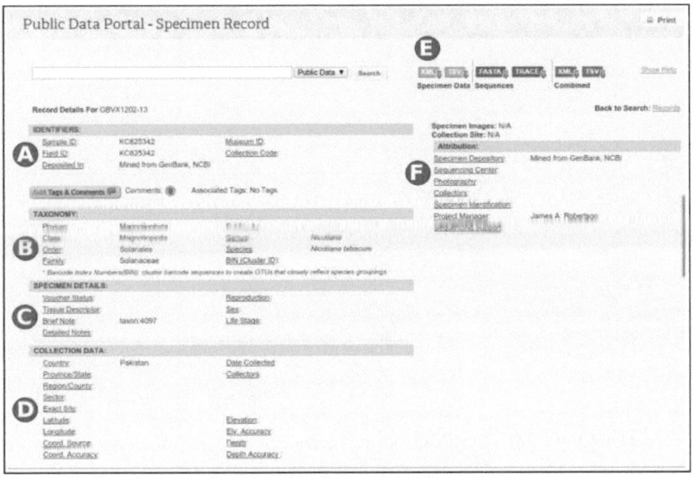

Figure 19: Specimen registration details in *BOLD* (Part 1)

Figure 19 (A) shows the species record identifiers, unique database access codes. Marker B of figure 19 shows the species' complete taxonomy, followed by more detailed information on the species (C) and information on the location where the data was collected (D).

BOLD also makes it possible to acquire species (E) data in formats such as XML (Extensible Markup Language) and TSV (Tab-separated values) - tabular data that can be accessed using spreadsheet editors. Sequence data can be downloaded in FASTA and TRACE format.

In the last block (F), still in figure 19, information on authorship and where the sample data is stored is shown. The rest of the species record page can be seen in figure 20. This part contains *DNA* information

Figura 20: Details of the specimen record in *BOLD* (Part 2) that served as the search key in the database, the *DNA* sequence, the amino acids it encodes and an *"Illustrative barcode"* - the species' *DNA barcode*.

The *DNA* barcodes displayed in the *BOLD* records are similar to the barcodes on supermarket products, but the bars are uniform in size and have four distinct colours: red, black, blue and green, representing thymine, guanine, cytosine and adenine respectively. The barcode can be different sizes as it depends on the number of *DNA* base pairs used as a key to represent the species. In species where the *matK* gene is also available, a table similar to the one shown in figure 20 is displayed, but with the *DNA* and amino acid sequence and the *barcode* of the *matK* gene.

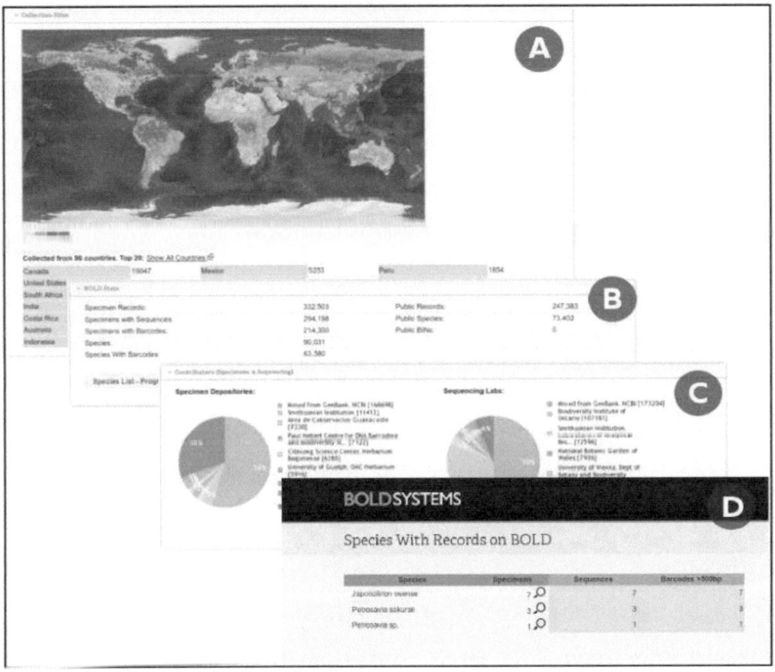

Figure 21: *BOLD*'s geolocalisation and statistics pages.

BOLD offers various statistics on the database and the geographical location of the specimens (figure 21). It is clear which countries the samples were collected in (A), which research centres and species repositories have provided information (C), the number of specimens in the database, the number of specimens with *DNA barcode* records (B) and a specific table on the species and its stage of sequencing (D).

The BOLD system has facilitated the evolution of *DNA barcode* research, mainly by providing access to genomic data obtained by the community

and providing researchers with tools to manage this data.

CHAPTER 3

APPLICATIONS AND IMPORTANCE OF GENOME-BASED CLASSIFICATION AND IDENTIFICATION SYSTEMS FOR LIVING BEINGS

Since the period of research carried out by Linnaeus and his predecessors, the classification and naming of living beings has undergone constant transformation with the aim of making these systems increasingly efficient and effective (PRESTES, 2009). The concern at the time of Dioscorides to classify plants for medical purposes can still be considered a contemporary concern. Plants used throughout human history for curative purposes are still in evidence, subsidising current popular medicine (BLASILEIRO, 2008).

Despite the varied, current and promising possibilities for species identification, accessibility to these resources is still scarce, so that most taxonomists still have to resort to so-called dichotomous keys.

Dichotomous keys present a list of characteristics (grouped in pairs) which are usually related to the visible characters of living beings. At the end of each characteristic, a code is inserted (usually a number or letter) which leads to another set of characteristics. The process is repeated until the name, genus or family of the species being researched is determined.

1) a) .... Palm tree2
b) Trees or shrubs3
2) a) Above-ground stem .... *Syagruspetraea*
b) Underground stem .... *Attaleageraenses*
3) a) Limb number more than 14
b)Number of limbs 129
4) a) Digitised leaflet arrangement5
b) Pinnate or bipinnate leaf arrangement9
5) a) Alternating phyllotaxis and retuse limbal apex .... *Eriothecagracilipes*
b) Opposite or subopposite phyllotaxis6

Figure 22: Fragment of the dichotomous key to tree and shrub species.

Source: Adapted from Mahmoud (2003)

It is possible to find several works creating or revising dichotomous keys for various animal and plant groups, examples of which can be seen in Mahmoud *et al.* (2003), Garcia and Odebrecht (2009) and Silva *et al.* (2011).

The fragment of the dichotomous key created by Mahmoud *et al.* (2003), shown in Figure 22, shows the use of morphological characteristics, the basis of this form of identification.

Keys are useful for identifying living beings, but they are still based on one-off efforts. Researchers usually create identification keys for a small group of beings, usually families or genera from a particular location, restricting the use of this key (SILVA, 2009).

In some branches of science, such as phytochemistry, pharmacological research and even the food industry, the identification of living beings is fundamental (MACIEL, 2002). For the pharmaceutical industry and ethnopharmacological studies, for example, more accurate identification techniques are indispensable, as the misidentification of species can lead to serious health problems for the population for whom the pharmaceutical products or drugs produced from plants are intended (CAVALCANTI & ALBUQUERQUE, 2013).

Identification using dichotomous keys is often insufficient, especially for plant species (see AMORIM, 2006). The limitations of this method often prevent identification, which can only be completed using other techniques or with the help of a specialist in the field of study.

One of the most promising current techniques is the use of genomes to identify species. In this sense, the following subtopics will deal with the use of molecular techniques based on genome analysis to identify pathogenic bacterial species (section 3.1), tree species (section 3.2) and medicinal plants (section 3.3).

3.1 IDENTIFICATION OF PATHOGENIC BACTERIA

DNA analysis has also been used in the diagnosis of pathogenic bacteria in order to obtain more precise correlations between clinical isolates and epidemiological investigation records.

Bacteria pose a risk in the hospital environment, as they can trigger outbreaks of hospital-acquired infections. Associated with this imminent danger is the fact that some of these microorganisms have characteristics that make conventional identification difficult or even impossible (HARRIS & HARTLEY, 2003).

Knowing the microorganism causing the infection is crucial for adopting control measures, which usually involve the use of antibiotics. Delays or errors in the identification process can lead to a worsening of the patient's clinical condition or the hospital infection (COIMBRA, 2011).

In this sense, one of the main technologies is identification through rDNA sequencing, since sequencing associated with sequence alignment involves low risk in the process of analysing and interpreting the results (CLOUD *et al.* 2002). In this way, they produce a lower probability of false negatives and false positives.

By sequencing the rDNA fragment and analysing it using tools such as BLAST against a database of rDNA sequences, the pathogen can be accurately identified.

In a study carried out at the Emergency Unit in Alagoas, Brazil, Coimbra (2011) was able to identify around 88% of the species of pathogenic bacteria in a sample of 202 microorganisms by analysing the 16S gene. The molecular technique for identifying bacteria is particularly useful in cases where bacteria cannot be identified phenotypically.

3.2 IDENTIFYING TREE SPECIES TO CONTROL DEFORESTATION

Molecular analyses using DNA are being used successfully to monitor and control deforestation. Various studies (KRESS, 2005; GONZALEZ, 2009; LOWE *et al.,* 2011; NITHANIYAL *et al.,* 2014) have already pointed to the possibility of identifying wood, including that which is often traded illegally.

According to Nithaniyal *et al.* (2014) in a study involving species in India, it was possible to correctly identify 136 tree species out of 143 surveyed (95%) using the *rbcL* and *matK* genes. The work was carried out by extracting DNA samples and consulting the *DNA barcode* database. The power of this method to discriminate between species makes it easier to monitor this important natural resource, as well as helping to formulate better biodiversity conservation plans.

Thompson and Newmaster (2014) pointed out that research using *DNA barcoding* in 335 forest plots in northeastern Ontario, Canada, provided more accurate estimates (42% more species identified) than traditional morphological taxonomic survey methods, and also concluded that identification using traditional identification methods was 37% more expensive than identification using *DNA barcoding*.

The next section of this paper (3.3) looks in more detail at the recurring problems associated with the identification of medicinal plants and the solutions provided by molecular analyses.

3.3 THE CASE OF MEDICINAL PLANTS

Today, more and more people are interested in getting to know and using medicinal plants, either because they are looking for a more natural and healthy life or because of financial difficulties faced with the high prices of industrialised medicines (QUEIROZ, 2012). Street markets and natural products shops in large urban centres have provided a wide variety of medicinal herb products.

In Brazil, the medicinal plants consumed have little or no evidence of their pharmacological properties (VEIGA *et aL,* 2005). Neglecting the effects of a particular herb can lead to problems ranging from the absence of the expected healing properties to serious poisoning, which can lead to death.

Veiga *et al.* (2005) discuss some of the problems that can occur with the commercialisation of medicinal plants, such as: (i) misidentification; (ii) the possibility of adulteration in extracts, in the case of the commercialisation of plant powder, usually contained in capsules, bottles or sachets; (iii) interactions between medicinal plants and allopathic medicines; (iv) overdose effects, as well as (v) allergic or toxic reactions.

Dichotomous keys are a commonly used tool for identifying plant species, and when they are not enough, the help of taxonomic specialists can be enlisted.

However, considering the use of dichotomous identification keys, there are problems related to access to the reproductive parts of the species to be identified. Some keys that require knowledge of the floral characters of plant species, for example, may have their use restricted to the flowering period of these plants (VEIGA and PINTO, 2002). When young, tree species may have different textures on their stem coatings or few or no leaves during a given season.

Figure 23: Powder from the bark of a herbal plant
Source: Centroflora Group[7]

In order for science to make use of popular knowledge, a rigorous identification process is necessary, especially with regard to plants, which are more numerous and show greater variety. Procópio and Secco (2008) highlight the importance of identifying plant species for monitoring and drawing up forest management plans.

The recurring problem when the object of study is plants is the fact that many products with healing properties are marketed in fragments (Figure 23), powders or even whole parts of the plant, but which do not provide the possibility of identification through dichotomous keys or even through the evaluation of a taxonomic expert.

The difficulty of identifying species using inadequate methods can lead both researchers and users of medicinal plants into error, since in many groups of living beings the morphological differences are small, and one or more characters may not be visible (MARTINS-DA-SILVA, 2002).

The use of identification keys, despite its limitations, is still an easy and economical way to identify species and should not be discarded as a tool to aid research.

The gaps in identification methods, whether through the use of dichotomous keys

[7]Available at: <http://www.centroflora.com.br/criacao-de- produto-tradicional- fitoterapico-tera- consulta publica/> Accessed on: 21 Aug. 2015.

or expert analysis, can be remedied with the 7 Available at: <http://www.centroflora.com.br/criacao-de- produto-tradicional- fitoterapico-tera- consulta-publica/> Accessed on: 21 Aug. 2015.

the use of *DNA barcoding,* since there is the possibility of extracting DNA and sequencing samples in the form of small fragments.

In this sense, the powdered extract of a particular plant species, which is difficult to identify, can be analysed using its genetic material, which, when sequenced, can be consulted in a database and the corresponding species returned as a result.

FINAL CONSIDERATIONS

The remarkable evolution of classification systems over the centuries has continually provided new challenges. The first classifications that met the needs of Aristotle's time became insufficient over the decades and new solutions had to be created.

When anatomical study is insufficient, other strategies are adopted, such as in the case of beings from the monera kingdom, biochemical analyses or genome analyses are used to identify the species. Tests, especially molecular tests that make it possible to identify bacteria using DNA, have evolved rapidly, partly because of the need to identify pathogenic bacterial agents, an indispensable task in hospital infection control.

The determination of genes that are conducive to distinguishing and classifying living beings is still the subject of debate and constant study. However, the most recent publications that accredit the *COI* genes for animals*, matK* and *rbcL* in the *Plantae* kingdom*, ITS* for fungi and some intergenic stretches in the Monera kingdom, have already provided good results, such as those observed in HAJIBABAEI (2006), MEUSNIER (2008) and SCHOCH (2012).

BOLD provides a means for researchers to identify animals, plants and fungi through segments of their genomes. It also serves as a platform for registering specimen data, as well as providing tools for manipulating genetic data, serving as a reference for research into living beings around the world.

The *DNA barcode*, which for the time being appears as an *"illustrative barcode"*, is not yet effectively used, like the barcodes on market products, and the manual sequencer mentioned in Pennisi (2007) has not yet been developed, but the possibility of identifying living beings using short stretches of DNA could produce a real revolution.

The application of *DNA Barcoding* in areas such as botany, for example, would solve the problems related to plant identification, since the variety of species or the way they are presented can prevent precise identification. The same process that identifies can be used to discover new species, by associating the DNA of the species under investigation with the numerous DNAs already registered in public databases and, by analysing the divergences, the fact that it is a new organism can be considered.

Although there is still a lot of controversy about the study of genes in relation to phylogeny, it is known that it is possible to make evolutionary associations between species using the genome. The work of Zhang *et al.* (2001) with groups of bacteria, for example, suggested the transfer of a change in the taxonomy of some species.

The Angiosperm Phylogeny Group (APG) has also achieved significant results in its studies of phylogenies in angiosperm groups. In their work, *APG* also presented the emergence of new orders and a phylogeny that diverges from those commonly adopted in analyses that did not take genes into account.

Countless efforts have been and still are being made to consolidate the best genes that can correctly discriminate and classify species. The collaborative aspect around BOLD creates a task force capable of providing a consistent research platform that is revolutionising the way we see life on the planet.

REFERENCES

ABREU, Y, S. Aristotle's method for the study of living beings. **Revista da SBHC.** n. 11.1994.

ACKERFIELD, J.; WEN, J. Evolution of Hedera (the ivy genus, Araliaceae): insights from chloroplast DNA data. **International Journal of Plant Sciences,** v. 164, n. 4, p. 593-602, 2003.

ALTSCHUL, Stephen F. *et al.* Basic local alignment search tool. **Journal of molecular biology,** v. 215, n. 3, p. 403-410, 1990.

AMORIM, Isaac Lucena de *et al.* Morphological aspects of Trema seedlings and saplings. **Brazilian Seed Journal,** 2006.

BERGTHORSSON, U. *et al.* Widespread horizontal transfer of mitochondrial genes in flowering plants. **Nature,** v. 424, n. 6945, p. 197-201, 2003.

BLAXTER, M. L. The promise of a DNAtaxonomy. **Philosophical Transactions of the Royal Society of London B: Biological Sciences,** v. 359, n. 1444, p. 669-679, 2004.

BLUNT, W. **Linnaeus:** The Compleat Naturalist. French Lincoln Limited. London. 2004. Available at: <http://books.google.com.br/books?id=FRH_EMhQYhYC&printsec=frontcover&dq=Linnaeus&hl=en-BR&sa=X&ei=vBmHT_3BA8el8AGm06iyCA&ved=0CDIQ6AEwAA#v=onepage&q=Linnaeus&f=true>. Accessed 17 Mar. 2015.

BOLD. Data Release Packages. Available at: <http://www.boldsystems.org/index.php/datarelease>. Accessed on: *TJ* Feb. 2015.

BRASILEIRO, B. G.; PIZZIOLO, V. R.; MATOS, D. S.; GERMANO, A. M.; JAMAL, C. M. Medicinal plants used by the population assisted in the "Family Health Programme", Governador Valadares, MG, Brazil. **Brazilian Journal of Pharmaceutical Sciences,** v. 44, n. 4, 2008.

BREMER, B. *et al.* An ordinal classification for the families of flowering plants. **Ann. Missouri Bot. Gard.** v. 85 p. 531-553, 1998.

BREMER, B. *et al.* An update of the Angiosperm Phylogeny Group classification for the orders and families of flowering plants: APG II. **Bot J Linn Soc,** v. 141, p. 399-436, 2003.

BREMER, B. *et al.* An update of the Angiosperm Phylogeny Group classification for the orders and families of flowering plants: APG III. **Botanical Journal of the Linnean Society,** 2009.

BRZOZOWSKI, J. A. Is it possible to place biological species in cladograms? **Acta Scientiae,** v. 16, n. 2, 2014.

BRZOZOWSKI, J. A. **Biological taxa: semantic and metaphysical aspects.** Florianópolis, SC. Originally presented as a doctoral thesis, Federal University of Santa Catarina, 2012.

CALDEIRA, A. M. A.; ARAÚJO, E. S. N. N., Introdução à didática da biologia, Escrituras Editora e Distribuidora de Livros Ltda. São Paulo - SP, 2009. Available at <https://books.google.com.br/books?id=15ArCAAAQBAJ>. Accessed on 26 July 2016.

CAVALCANTI, R. C.; ALBUQUERQUE, U. P. The "Hidden Diversity" of Medicinal Plants in Northeastern Brazil: Diagnosis and Prospects for Conservation and Biological Prospecting. **Evidence-Based Complementary and Alternative Medicine.** Hindawi Publishing Corporation: New York. 2013. Available at: <http://www.hindawi.com/journals/ecam/2013/102714/>. Accessed on: 10 January 2014.

COIMBRA, D. G. Identification of pathogenic bacteria isolated in the Agreste-AL Emergency Unit, through broad-spectrum PCR and 16S rRNA sequencing. 2011.

CLOUD, J. L.; NEAL, H.; ROSENBERRY, R.; TURENNE, C. Y; JAMA, M; HILLYARD, D. R.; CARROLL, K. C. Identification of Mycobacterium spp. by using a commercial 16S ribosomal DNA sequencing kit and additional sequencing libraries. **Journal of clinical microbiology,** v. 40, n. 2, p. 400-406, 2002.

COUTINHO, D. E; TRAVASSOS, L. M. A.; DO AMARAL, F. M. M. Ethnobotanical study of medicinal plants used in indigenous communities in the State of Maranhão-Brazil. **Visão Acadêmica,** v. 3, n. 1, 2002.

DE OLIVEIRA, J. A. *et al.* Mitochondrial gene sequences for animal species identification. **Colombian Journal of Animal Science,** v. 5, n. 2, p. 396-407, 2013.

DOYLE, J. J.; GAUT, B. S. Evolution of genes and taxa: a primer. **Springer Netherlands.** ImPLANT MOLECULAR EVOLUTION, p. 1-23, 2000.

ECK, R. V; DAYHOFF, M. O. BARKER, W. C.; DAYHOFF, M. O.; DAYHOFF, M. O. Atlas of protein sequence and structure. **Atlas of Protein Sequence and Structure,** v. 5, p. 31-39, 1972.

EDGAR, Robert C. MUSCLE: multiple sequence alignment with high accuracy and high throughput. **Nucleic acids research,** v. 32, n. 5, p. 1792-1797, 2004.

FELSENSTEIN, J. Evolutionary trees from DNA sequences: a maximum likelihood approach. **Journal of molecular evolution,** v. 17, n. 6, p. 368-376,1981.

FOLMER, O.; BLACK, M.; HOEH, W.; LUTZ, R.; VRIJENHOEK, R. DNAprimers for amplification of mitochondrial cyto- chrome c oxidase subunit I from diverse metazoan invert- ebrates. **Mol. Mar. Biol. Biotechnol.,** 294-299, 1994.

FRANGSMYR, T.; HEILBRON, J. L. **The quantifying spirit in the 18th century.** Univ of California Press, 1990.

FREEMAN. S; HERRON J. C. **Evolutionary Analysis.** Artmed Editora, 2009. Available at: <https://books.google.com.br/books?id=JhZb-cuEhloC>. Accessed on 27 July.

2016.

GARCIA, M.; ODEBRECHT, C. Illustrated dichotomous key for the identification of Thalassiosira Cleve (diatom) species in the Patos Lagoon estuary and adjacent coastal area (Rio Grande do Sul, Brazil). **Biota Neotrop.** v. 9, n. 2. 2009.

GONZALEZ, M. A. *et al.* Identification of Amazonian trees with DNA barcodes. **PLoS one,** v. 4, n. 10, p. e7483, 2009.

GROUP, CBOL Plant Working *et al.* A DNA barcode for land **plants.Proceedings of the National Academy of Sciences,** v. 106, n. 31, p. 12794-12797, 2009.

HAJIBABAEI, M. *et al.* DNA barcodes distinguish species of tropical Lepidoptera. **Proceedings of the National Academy of Sciences of the United States of America,** v. 103, n. 4, p. 968-971, 2006a.

HAJIBABAEI, M.; SMITH, M.; JANZEN, D. H.; RODRIGUEZ, J. J.; WHITFIELD, J. B.; HEBERT, P. D. A minimalist barcode can identify a specimen whose DNA is degraded. **Molecular Ecology Notes,** v. 6, n. 4, p. 959-964, 2006.

HAMILTON, A.C. Medicinal plants, conservation and livelihoods. **Biodiversity and Conservation** 13: 1477-1517, 2004.

HARRIS, K. A.; HARTLEY, J. C. Development of broad-range 16S rDNA PCR for use in the routine diagnostic clinicai microbiology Service. **Journal of medical microbiology,** v. 52, n. 8, p. 685-691, 2003.

HEBERT, P. D. N. *et al*. Biological identifications through DNA barcodes.

Proceedings of the Royal Society of London B: Biological Sciences, v. 270, no. 1512, p. 313-321, 2003.

HIGGINS, D. G.; SHARP, Paul M. CLUSTAL: a package for performing multiple sequence alignment on a microcomputer. **Gene,** v. 73, n. 1, p. 237-244,1988.

HOLLINGSWORTH, P. M. *et* a/. A DNA barcode for land **plants.Proceedings of the National Academy of Sciences,** v. 106, n. 31, p. 12794-12797, 2009.

JARDINE, N; SECOND, J. A.; SPARY, E. C. **Cultures of Natural History,** Cambridge University, 1996. Available at: <http://books.google.com.br/books?hl=en-BR&lr=&id=z9_Z9UbUIeQC&oi=fnd&pg=PA145&dq=Linnaeus+the+botanist&ots=opejeYsX3v&sig=DI-Z6QlksfhYu_EWUliOpPlu95k#v=onepage&q=Linnaeus%20the%20botanist&f=true>. Accessed on: 17 March 2015.

JO, H.; GIM, J. A.; JEONG, K. S.; KIM, H. S.; JOO, G. J. Application of DNA barcoding for identification of freshwater carnivorous fish diets: Is number of prey items dependent on size class for Micropterus salmoides? **Ecology and evolution,** v. 4, n. 2, p. 219-229, 2014.

KRESS, W. J.; WURDACK, K. J.; ZIMMER, E. A.; WEIGT, L. A.; JANZEN, D. H. Use of DNA barcodes to identify flowering plants. **Proceedings of the National Academy of Sciences of the United States of America,** v. 102, n. 23, p. 8369-8374, 2005.

KUMAR, S.; TAMURA, K.; NEI, M. MEGA3: integrated software for molecular evolutionary genetics analysis and sequence alignment. **Briefings in bioinformatics,** v. 5, n. 2, p. 150-163, 2004.

LYNCH, M.; JARRELL, P. E. Amethod for calibrating mol- ecular clocks and its application to animal mitochondrial DNA. **Genetics,** 1197-1208, 1993.

LORENZI, H.; MATOS, F.J.A. **Plantas Medicinais no Brasil:** nativas e exóticas. 2ª ed. Nova Odessa, Plantaram Institute, 2008.

LORENZI. J. O. Variation in the culinary quality of cassava roots. **Bragantia,** v. 53, n. 2, 1994.

LOWE, A. J.; CROSS, H. B. The Applicat ion of DNA methods to Timber Tracking and Origin Verificat ion. **IAWA Journal,** v. 32, n. 2, p. 251-262, 2011.

MACHADO, D. S. L. O.; SARZI, N. M.; OLIVEIRA, V. M. S. In Silico Analyses of Phylogenetic Relationships between Plants. **Proceedings of the International Teaching Salon,**

Research and Extension, v. 4. n. 4. 2012.

MACIEL, Maria Aparecida M. *et al.* Medicinal plants: the need for multidisciplinary studies. **Química nova,** v. 25, n. 3, p. 429-438, 2002.

MAGNANI, G. S.; DIDONET, C. M.; CRUZ, L. M.; PICHETH, C. E; PEDROSA, E O.; SOUZA, E. M. Diversity of endophytic bacteria in Brazilian sugarcane. **Genet Mol Res,** v. 9, n. 1, p. 250-258, 2010.

MAHMOUD, A. G. E.; VIRILLO, C. B.; RIBEIRO, D. B .; IKEMOTO, E.; HORNINK, G. G., RICARTE, J. D.; ALCANTARA, S. E; WATANABE, T. M. Dichotomous Key for the Identification of Cerrado Tree Species in the Municipality of Itirapina - São Paulo . Campinas, SP. Department of Botany, Institute of Biology, State University of Campinas. 2003.

MARANDINO, M. Science Museums, Collections and Education: necessary relationships.Revista **Museologia e Património,** v. 2, n. 2, p. 1-12, 2009.

MARTINEZ, S. T.; ALMEIDA, M. R.; PINTO, A. C. Natural hallucinogens: a flight from Medieval Europe to Brazil. **Chem. Nova,** v. 32, n. 9, p. 2501-2507, 2009.

MARTINS-DA-SILVA, R. C. V. Identification of botanical specimens. Embrapa Amazônia Oriental, Belém-PA.v. 99, 2002 Available at: <http://pt.calameo.com/read/002327569f932a91767>. Accessed on: 30 July 2016.

MEDEIROS, C. A. M., BOICA-JUNIOR, A. L. TORRES, A. L. Effect of aqueous plant extracts on the oviposition of the cruciferous moth in cabbage. **Bragantia,** v. 64, n. 2, p. 227-232, 2005.

MEUSNIER, L; SINGER, G. A.; LANDRY, J. E; HICKEY, D. A.; HEBERT, P. D.; HAJIBABAEI, M. A universal DNA mini-barcode for biodiversity analysis. **BMC genomies,** v. 9, n. 1, p. 214, 2008.

NASCIMENTO, A. E J.; DE SOUZA, D. C. A look at the study of living beings in the middle ages: fundamental themes of biology in the philosophy of nature. **Theoria - Revista Eletrónica de Filosofia,** v. 3, n. 6, 2011.

NITHANIYAL, S. *et al.* DNA barcode authentication of wood samples of threatened and commercial timber trees within the tropical dry evergreen forest of india. **PloS one,** v. 9, n. 9, p. e107669, 2014.

PANTOJA, S. Taxonomy of Cryptogams. Castelo Branco University. Rio de Janeiro, RJ. 2007.

PENNISI, E. Proposed Plant. Wanted: a barcode for plants. **Science,** v. 318, n. 5848, p. 190, 2007.

PHILIPPE, H.; LARTILLOT, N.; BRINKMANN, H. Multigene analyses of bilaterian animals corroborate the monophyly of Ecdysozoa, Lophotrochozoa, and Protostomia. **Molecular biology and evolution,** v. 22, n. 5, p. 1246-1253, 2005.

PRESTES, M. E. B.; OLIVEIRA, P; JENSEN, G. M. The origins of Cari von Linné's classification of plants in biology teaching. **Philosophy and History of Biology,** v. 4, n. 1, p. 101-137, 2009.

PROCÓPIO, L. C.; SECCO, R. S. The importance of botanical identification in forest inventories: the example of "tauari" (Couratari spp. and Cariniana spp.- Lecythidaceae) in two managed areas in the state of Pará. **Acta Amazonica,** v. 38, n. 1, p. 31-44, 2008.

QUEIROZ, A. A. Cultivation of medicinal plants in public schools in the municipality of Seropédica. Course Conclusion Work (Specialisation) - Institute of Drug Technology. **Fiocruz Institutional Repository.** Rio de Janeiro, RJ, 2012.

RATNASINGHAM, S.; HEBERT, P. D. N. BOLD: The Barcode of Life Data System (http://www. barcodinglife. Org). **Molecular ecology notes,** v. 7, n. 3, p. 355-364, 2007.

RAVEN, P. H.; BERLIN, B.; BREEDLOVE, D. E. The origins of taxonomy. **Science,** v. 174, n. 4015, p. 1210-1213, 1971.

RESCH, M. C. *et al.* Where taxonomy based on subtle morphological differences is perfectly mirrored by huge genetic distances: DNA barcoding in protura (hexapoda). **PloS one,** v. 9, n. 3, p. E90653, 2014.

RIDLEY, M. **Evolution.** Artmed Editora, Porto Alegre - RS, 3ª .ed 2009.

RODRIGUES, M. H. A study of the influence of mysticism in the History of Science. **Impulso,** v. 25, n. 64, p. 93-106, 2016.

RODRIGUES, V. M. S. Using natural paints in the classroom from regional pigments and binders. Course Conclusion Work (Graduation) - University of Brasília, 2013.

RODRIGUES. W. A. Revision of botanical nomenclature: Important dates in the history of post-Linnean botanical nomenclature. **Publicatio UEPG: Biological and Health Sciences,** v. 9, n. 3, 2009.

ROKAS, A; KRUEGER, D; CARROLL, S. B. Animal evolution and the molecular signature of radiations compressed in time. **Science,** v. 310, n. 5756, p. 1933-1938, 2005.

RUPPENTHAL, R. D.; PEREIRA, F. D. S.; CANTARELLI, V. V.; SCHRANK, I. S. Application of broad-range bacterial PCR amplification and direct sequencing on the diagnosis of neonatal sepsis. **Braz. J. MicrobioL,** São Paulo , v. 36, n. 1, 2005 . Available at: <http://www.scielo.br/scielo.php?script=sci_arttext&pid=S1517-83822005000100006&lng=en&nrm=iso>. Accessed on: 18 Jun 2014.

RUSSO, C. de; MATIOLI, S. R. How to choose genes for specific phylogenetic problems. **Molecular biology and evolution.** Ribeirão Preto: Holos Editora, p. 130-136, 2001.

RZHETSKY, A.; NEI, M. Theoretical foundation of the minimum-evolution method of phylogenetic inference. **Molecular Biology and Evolution,** v. 10, n. 5, p. 1073-1095, 1993.

SACCONE, C.; DE CARLA, G.; GISSI, C.; PESOLE, G.; REYNES, A. Evolutionary genomics in the Metazoa: the mitochondrial DNA as a model system. **Gene,** 195-210,1999.

SAITOH, T. *et al.* DNAbarcoding reveals 24 distinct lineages as ciyptic bird species candidates in and around the Japanese Archipelago. **Molecular ecology resources,** v. 15, n. 1, p. 177-186, 2015.

SAITOU, N.; NEI, M. The neighbor-joining method: a new method for reconstructing phylogenetic trees. **Molecular biology and evolution,** v. 4, n. 4, p. 406-425, 1987.

SANTOS, C. M. D.; CALOR, A. R. Teaching Evolutionary Biology using the conceptual framework of Phylogenetic Systematics-I. **Ciência & Ensino (ISSN 1980- 8631),** v. 1, n. 2, 2008.

SANTOS, C. M. D.; KLASSA, B. Hennigian phylogenetic systematics: revolution or change within a paradigm? **Scientiae Studia,** v. 10, n. 3, p. 593-612, 2012.

SCHOCH, C. L. *et al.* Nuclear ribosomal internal transcribed spacer (ITS) region as a universal DNA barcode marker for Fungi. **Proceedings of the National Academy of Sciences,** v. 109, n. 16, p. 6241-6246, 2012.

SHAW, J. *et al.* The tortoise and the hare II: relative utility of 21 noncoding chloroplast DNA sequences for phylogenetic analysis. **American Journal of botany,** v. 92, n. 1, p. 142-166, 2005.

SILVA, B. D. **Taxonomic studies of species of the genus Culex (Diptera: Culicidae) from the Neotropical region, using subunit I of the mitochondrial cytochrome oxidase gene.** 2009. PhD Thesis. University of São Paulo. School of Public Health. Department of Epidemiology.

SILVA, P. G.; VAZ-DE-MELLO F. Z.; DI MARE, R. A. Identification guide to the species of Scarabaeinae (Coleoptera: Scarabaeidae) from the municipality of Santa Maria, Rio Grande do Sul, Brazil. **Biota Neotrop.** v. 11. n.4. 2011.

SINGH V.; PANDE, P. C.; JAIN, D.K., **A text of botany:** Angiosperms. Rastogi Publications. 2008. Available at: <http://books.google.com.br/books?id=knLkmo7CtYcC> Accessed on: 17 Mar. 2015.

SMITH, T. E; WATERMAN, M. I. S. Identification of common molecular subsequences. Journal of molecular biology. **Academic Press,** v.147, p 195-197, 1981.

SNEATH, P. H. A.; SOKAL, R. R. Unweighted pair group method with arithmetic mean. **Numerical Taxonomy,** p. 230-234,1973.

TABERLET, P.; COISSAC, E.; POMPANON, E; GIELLY, L.; MIQUEL, C.; VALENTINI, A.; VERMAT, T.; CORTHIER, G.; BROCHMANN, C.; WILLERSLEV, E. Power and limitations of the chloroplast trnL (UAA) intron for plant DNA barcoding. **Nucleic Acids Research,** v. 35, n. 3, p. el4-el4, 2007.

TAMURA, K.; PETERSON, D.; PETERSON, N.; STECHER, G.; NEI, M.; KUMAR, S. MEGA 5: molecular evolutionary genetics analysis using maximum likelihood, evolutionary distance, and maximum parsimony methods. **Molecular biology and evolution,** v. 28, n. 10, p. 2731-2739, 2011.

TATENO, Y; TAKEZAKI, N.; NEI, M. Relative efficiencies of the maximum-likelihood, neighbour-joining, and maximum-parsimony methods when substitution rate varies with **site.Molecular Biology and Evolution,** v. 11, n. 2, p. 261-277, 1994.

THOMPSON, K. A.; NEWMASTER, S. G. Molecular taxonomic tools provide more accurate estimates of species richness at less cost than traditional morphology-based taxonomic practices in a vegetation **survey.Biodiversity and conservation,** v. 23, n. 6, p. 1411-1424, 2014.

VEIGA. V. F. J.; PINTO, A. C. O GÊNERO Copaifera L. **Quim. Nova**, v. 25, n. 2, p. 273-286, 2002.

VEIGA, V. F. J.; PINTO, A. C.; MACIEL, M. A. M. Plantas medicinais: cura segura. **Química nova,** v. 28, n. 3, p. 519-528, 2005.

VEIGA, V. F. J. Study of the consumption of medicinal plants in the Centre-North Region of the State of Rio de Janeiro: acceptance by health professionals and mode of use by the population. **Rev bras farmacogn,** v. 18, n. 2, p. 308-13, 2008.

VON MURALT, M. The tree that became a country. **Revista USP,** n. 71, p. 171-198, 2006.

WILLIAMS, R. L. Botanophilian in Eighteenth-Century France. **The Spirit of the Enlightenment.** Archives internationeales d'histoire des idees, Dordrecht: kluwer academic, 2001.

YAMASHIRO, T.; FUKUDA, T.; YOKOYAMA, J.; MAKI, M. Molecular phylogeny of Vincetoxicum (Apocynaceae-Asclepiadoideae) based on the nucleotide sequences of cpDNA and nrDNA. **Molecular Phylogenetics and Evolution,** v. 31, n. 2, p. 689- 700, 2004.

YANG, Z. PAML: a program package for phylogenetic analysis by maximum likelihood. **Computer applications in the biosciences: CABIOS,** v. 13, n. 5, p. 555-556, 1997.

ZHANG, Z.; KUDO, T.; NAKAJIMA, Y; WANG, Y. Clarification of the relationship between the members of the family Thermomonosporaceae on the basis of 16S rDNA, 16S-23S rRNA internal transcribed spacer and 23S rDNA sequences and chemotaxonomic analyses. **International Journal of Systematic and Evolutionary Microbiology,** v. 51, n. 2, p. 373-383, 2001.

APPENDIX A

Figure 24 shows the diagram created by Ramom Llull in 1305 and published in 1512, entitled: *The Ladder of* Ascent *and Descent of the Intellect*. The image shows a hierarchy represented by the rungs of a ladder, and on the right, the images illustrating the groups: Lapis

Figure 24: The ladder of ascent and descent of the intellect.[8]

(stone), Flamma (fire), planta (plant), Brutum (beast), Homo (human), Caelum (sky), Angelus (angel) and Deus (God).

[8] *Available at* <https://www.brainpickings.org/2012/05/29/trees-of-life-a-visual-history-of-evolution/>; Accessed on: 10 Jul. 2016

APPENDIX B

This appendix shows phylogenetic trees created with MEGA software, version 6.0, from sequences corresponding to the *rbcL* genes of *Zea mays, Nicotiana tabacum, Pinus thunbergii* from the *Plantae* kingdom *and Thiobacilus denitrificans*, a bacterium, obtained by consulting the NCBI Genbank database. The following methods were used: *Maximum Parsimony* (figure 25), *Neighborn-Joining* (figure 26), *Minimum Evolution* (figure 27), *Maximum likelihood* (figure 28) and UPGMA (figure 29), in all cases using the *software*'s default parameter values.

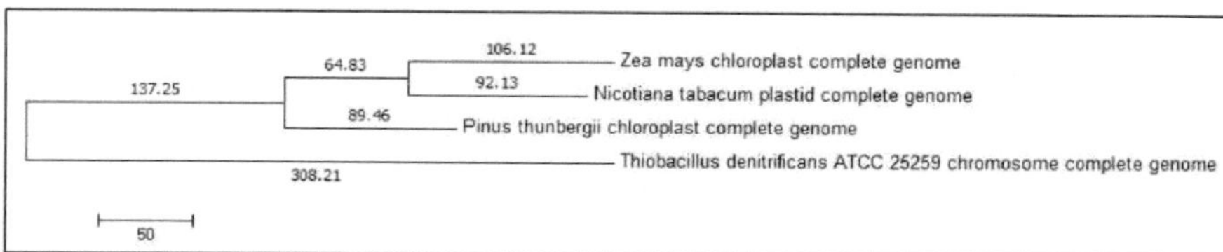

Figure 25: Phylogenetic reconstruction using *Maximum Parsimony*

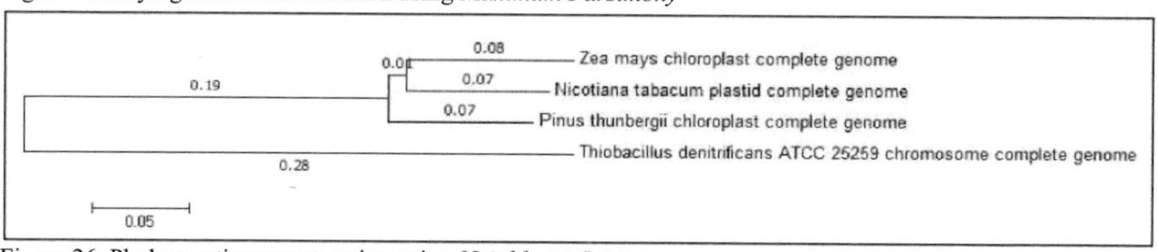

Figure 26: Phylogenetic reconstruction using *Neighborn-Joinig*

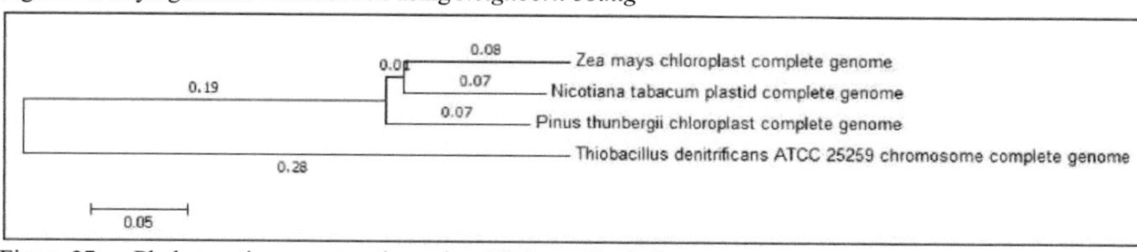

Figura 27: Phylogenetic reconstruction using *Minimum Evolution*

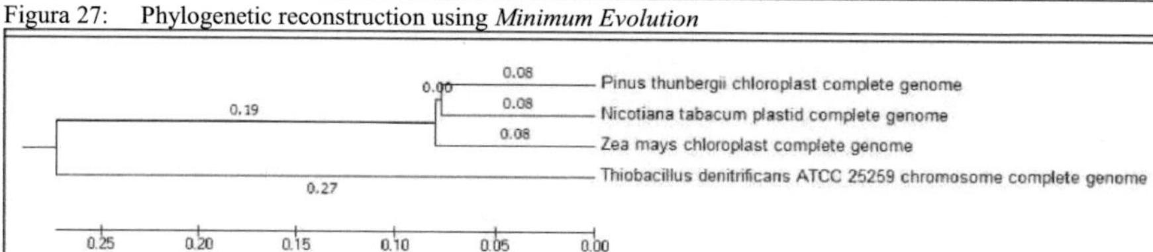

Figure 29: Phylogenetic reconstruction using UPGMA

All the methodologies used to generate phylogenetic trees, despite the small sample size, showed very similar results, with only small differences in the divergence time between the species' genes. In the case of the *Neighbom-Joining* and

Minimum Evolution methodologies, the trees produced were identical.

MIX
Papier aus verantwortungsvollen Quellen
Paper from responsible sources
FSC® C105338

Printed by Books on Demand GmbH, Norderstedt / Germany